Characterization and Manufacturing of Nano-Composites and Nano-Composite Coatings

Characterization and Manufacturing of Nano-Composites and Nano-Composite Coatings

Guest Editor

Nikolaos E. Karkalos

Basel • Beijing • Wuhan • Barcelona • Belgrade • Novi Sad • Cluj • Manchester

Guest Editor
Nikolaos E. Karkalos
School of Mechanical
Engineering
National Technical University
of Athens
Zografou
Greece

Editorial Office
MDPI AG
Grosspeteranlage 5
4052 Basel, Switzerland

This is a reprint of the Special Issue, published open access by the journal *Coatings* (ISSN 2079-6412), freely accessible at: https://www.mdpi.com/journal/coatings/special_issues/nano_composites_coatings.

For citation purposes, cite each article independently as indicated on the article page online and as indicated below:

Lastname, A.A.; Lastname, B.B. Article Title. *Journal Name* **Year**, *Volume Number*, Page Range.

ISBN 978-3-7258-8155-0 (Hbk)
ISBN 978-3-7258-8156-7 (PDF)
https://doi.org/10.3390/books978-3-7258-8156-7

Contents

About the Editor

Nikolaos E. Karkalos

Nikolaos E. Karkalos is a postdoctoral researcher with the Laboratory of Manufacturing Technology, School of Mechanical Engineering, NTUA, Greece. His research interests are related to the scientific field of computational methods for the simulation of systems in manufacturing, materials engineering and energy applications in the field of Mechanical Engineering using Machine Learning methods, the Finite Element Method (FEM), the Finite Volume Method (FVM), meshless methods and the Molecular Dynamics (MD) method.

Preface

This Reprint brings together recent research on nano-materials and nano-composite coatings for advanced engineering applications. Its subject covers the development, processing, characterization, and performance of coatings designed to improve wear resistance, corrosion protection, thermal stability, mechanical strength, and multifunctional surface behavior. The aim of this scientific work is to present a focused overview of current progress in the field and to show how nanoscale engineering can be used to create more durable and reliable surfaces for demanding service conditions. It is motivated by the growing need for high performance materials in sectors such as aerospace, automotive, energy, biomedical engineering, and manufacturing. This Reprint is addressed to researchers, engineers, and graduate students interested in surface engineering, materials science, and composite design, as well as to practitioners seeking recent developments with practical relevance.

Nikolaos E. Karkalos
Guest Editor

coatings

MDPI

Editorial

Editorial for Special Issue: "Characterization and Manufacturing of Nano-Composites and Nano-Composite Coatings"

Nikolaos E. Karkalos

Laboratory of Manufacturing Technology, School of Mechanical Engineering, National Technical University of Athens, 157 80 Athens, Greece; nkark@mail.ntua.gr

The continuous research progress in materials science has enabled the development of advanced nano-materials, including carbon nano-tubes, graphene and metal oxides with specialized properties, which can fundamentally affect the mechanical, thermal and tribological properties of conventional materials when used in the reinforcing phase [1,2]. In particular, nano-materials and nano-composites can form suitable materials for coatings in advanced applications due to their high mechanical strength, adhesion and higher durability compared to traditional coatings [2–4]. Thus, multifunctional surfaces with a combination of corrosion and wear resistance, anti-fouling properties and even self-healing capabilities can be created, considerably prolonging the lifespan and performance of mechanical components under demanding conditions [4–6].

Beyond their intrinsic material enhancements, nano-composites offer unique opportunities for tailoring surface behavior at the micro- and nano-scale, enabling precise control over a wide variety of related phenomena, such as friction, wear, corrosion, and thermal transport mechanisms [7,8]. Their ability to form dense, defect-resistant microstructures and to incorporate functional nano-particles makes them ideal for designing coatings with tunable properties that respond to external stimuli, such as temperature, mechanical loading, or chemical exposure [9,10]. As a result, nano-composite coatings are becoming increasingly common in sectors such as aerospace, the automotive industry, biomedical devices, and energy systems, where reliability, lightweight design, and long-term durability are critical [11,12]. This growing technological relevance highlights the need for continued research into the optimization of nano-composite synthesis procedures, interface analysis, and long-term performance evaluation under realistic operating conditions [13].

Given the significance of developing nano-composites and nano-composite coatings, this Special Issue collates high-quality research related to the most recent advancements in this scientific field, with applications spanning across various engineering disciplines, including materials science, mechanical engineering, and surface engineering [2]. The contributions cover novel synthesis routes, advanced characterization techniques, and innovative application-driven designs that highlight the multifunctional potential of nano-structured materials [1,5].Thus, these studies provide a comprehensive foundation for researchers and engineers exploring the integration of nano-composites and nano-composite coatings for advanced surface protection and multifunctionality in real-world applications. Each contribution to this Special Issue is briefly summarized below.

Liu et al. [14] investigated the effect of advanced coatings on enhancing the performance of a Wire Electrical Discharge Machining (WEDM) electrode for the machining of an Inconel 718 alloy, a material usually employed in aerospace applications due to its high strength and high temperature resistance. In their work, they applied the Taguchi-Data Envelopment Analysis-based Ranking (DEAR) methodology for the optimization of various process responses, such as the material removal rate, kerf width, and surface roughness. Their findings showed that the optimal parameters in this case were identified as the following: 140 μs pulse-on time (Ton), 50 μs pulse-off time (Toff), 60 V servo voltage (SV), and 5 kg wire tension (WT), achieving 1.1% error. Moreover, it was determined that among all tested parameters, pulse-off time was the dominant factor due to its crucial role in deionization during the process.

In another contribution, Liu et al. [15] studied the synthesis of lightweight, nano-porous Co-Ni/C composites using zeolitic imidazolate framework-67 (ZIF-67) as a template, involving $Ni(NO_3)_2$ etching and pyrolysis at 800 °C in an Ar/H_2 atmosphere with tunable morphology, which was achieved by varying Ni doping content to optimize microwave absorption properties through dielectric loss, magnetic loss, and component synergies. The results of this study demonstrated that the developed composites exhibit exceptional performance, achieving a maximum reflection loss (RL) of 72.88 dB and an effective bandwidth of 7.04 GHz (9.76–16.8 GHz) at a thickness of 2.53 mm when using 0.32 g $Ni(NO_3)_2$. Thus, these findings highlighted the potential of Co-Ni/C materials as high-efficiency broadband microwave absorbers for stealth applications, resisting electromagnetic interference.

The experimental study of Thangamani et al. [16] focused on the electrochemical micromachining (ECMM) of a corrosion-resistant Incoloy 825 alloy, particularly on the creation of micro-holes with silver-coated copper tool electrodes in sodium nitrate electrolyte. As well as exploring the effect of the coating, other parameters such as voltage, concentration, frequency, and duty cycle were optimized to improve material removal rate (MRR), overcut, conicity, and circularity errors. The results revealed that the silver-coated electrodes consistently outperformed uncoated ones, achieving higher MRR due to superior conductivity, reduced overcut and conicity of the holes and produced a cleaner surface topography with fewer cracks, pits, or embedded particles, as confirmed by SEM and EDS analyses. Finally, pulse frequency emerged as the key factor affecting MRR for uncoated tools while electrolyte concentration was the most important factor for coated ones. This study made several important recommendations, such as the use of coated electrodes with high voltage, the use of a high duty cycle and low frequencies for increased productivity, and the use of low voltage and high frequency for higher precision.

The study of Ashkinazi et al. [17] focused on the growth of a micro-nano crystalline diamond film on an axial carbide tool. The results showed that the temperature varied sharply along the cylindrical tool during diamond-coating growth, cooling most rapidly near the molybdenum holder, and that extending the free length or rotating the tool helps equalize heating. Moreover, rotation above 12 rpm enables uniform coating growth up to 40 mm and produces systematic grain-size variations that correlate with local surface temperature and distance from the plasma center. Without rotation, grain size decreases and diamond does not form on the underside, highlighting how controlled rotation and thermal management are essential for achieving consistent nano-crystalline diamond coatings on high aspect-ratio carbide tools.

In an experimental study, Kiiamov et al. [18] determined the size of iron oxide nano-particles using different methods. These experiments showed that the nano-particles have a three-layer structure consisting of a crystalline magnetic core, an amorphous paramagnetic shell, and an organic oleate coating. The measured particle sizes differ depending on the technique, i.e., TEM, XRD, and magnetic analysis, because only part of each particle is crystalline or magnetically active, while the organic shell does not contribute to crystallinity. The synthesis route enables controlled tuning of particle size and magnetic behavior for biomedical use, though any medical application will require rigorous validation.

 https://doi.org/10.3390/coatings16050548

Korzekwa et al. [19] showed that ultrasonic fragmentation and controlled deposition strongly influence the distribution of IF-WS_2 nano-particles within Al_2O_3 coatings. Among the two methods tested, the first method, without additional ultrasonic bath treatment, produced a far more uniform microstructure. Structural analyses confirmed the presence of WS_2- and WO_3-derived phases, while wettability tests showed that the first method yielded lower contact angles and higher surface free energy, indicating improved hydrophilicity due to better nano-particle dispersion. The study also proved that anodizing conditions and deposition methods can jointly affect surface behavior, laying the groundwork for future tribological testing and surface geometry evaluation.

Ibarra et al. [20] developed HVOF-sprayed coatings and found that adding 12wt.% NiCrFeSiBC + Mo to WC-CoCr significantly improves wear resistance, reducing the specific wear rate by about 30% at high loads while maintaining high hardness. The same 12% composition also leads to the greatest fracture toughness improvement and lowers friction by forming a more stable tribolayer during sliding. Both modified coatings (12% and 17%) reduce the coefficient of friction compared to pure WC-CoCr, but the 12% blend provides the best overall balance of toughness and wear performance.

Tuhta and Günday [21] demonstrated that applying a 1 mm MgO nano-coating to a steel chimney slightly increases its stiffness, reducing peak displacements from 150 mm to 140 mm and lowering maximum and minimum von Mises stresses at the base. The retrofit does not alter mode shapes but increases natural frequencies, indicating a stiffer structural system without adding significant mass which is a clear advantage over traditional strengthening methods. Thus, the study demonstrates that MgO nano-coatings can meaningfully improve the dynamic performance of industrial steel chimneys and hold strong potential as a lightweight, corrosion-resistant retrofitting technique.

Yuce and Tuhta [22] carried out a study on the structural behavior of a coated biogas digester under earthquake conditions. It was shown that a ZrO_2 nano-coating dramatically increased the stiffness of the PVC-PET biogas digester, increasing its natural frequencies and sharply reducing both displacements and von Mises stresses under seismic loading. Maximum displacement decreased by 77%, and peak stresses were reduced by more than 70%, showing that the coating effectively stabilizes the structure without adding significant mass. Thus, this study further confirmed that nano-coatings can greatly enhance the seismic performance and durability of flexible biogas digesters, offering a lightweight and corrosion-resistant alternative to traditional retrofitting methods.

The guest editor of this Special Issue would like to sincerely thank the authors for their high-quality contributions, the reviewers for their valuable work for the improvement of the quality of the submitted manuscripts, as well as the publisher for their work and cooperation.

Conflicts of Interest: The author declares no conflicts of interest.

References

1. Li, Y.; Liu, S.; Feng, F.; Li, Y.; Han, Y.; Tong, X.; Gao, X. Preparation and Characterization of Graphene Oxide/Carbon Nanotube/Polyaniline Composite and Conductive and Anticorrosive Properties of Its Waterborne Epoxy Composite Coatings. *Polymers* **2024**, *16*, 2641. [CrossRef]
2. Muresan, L.M. Nanocomposite Coatings for Anti-Corrosion properties of Metallic Substrates. *Materials* **2023**, *16*, 5092. [CrossRef]
3. Priyadharshini, R.; Xavier, J.R. Recent Advances in Graphene-Based Nanocomposite Coatings for Corrosion Resistance in the Automotive Industry and Their Smart Functional Applications. *J. Bio- Tribo-Corros.* **2025**, *11*, 113. [CrossRef]
4. Farooq, S.A.; Raina, A.; Mohan, S.; Singh, R.A.; Jayalakshmi, S.; Haq, M.I.U. Nanostructured Coatings: Review on Processing Techniques, Corrosion Behaviour and Tribological Performance. *Nanomaterials* **2022**, *12*, 1323. [CrossRef] [PubMed]
5. Vinodhini, S.P.; Xavier, J.R. Recent progress in graphene-based nanocomposites for enhanced energy storage and corrosion protection. *J. Mater. Sci.* **2025**, *60*, 14837–14879. [CrossRef]

https://doi.org//10.3390/coatings16050548

6. Aliofkhazraei, M. *Anti-Abrasive Nanocoatings*, 1st ed.; Woodhead Publishing: Cambridge, UK, 2014.
7. Wani, S.M.; Ahmad, B.; Saleem, S.S. Nano-mechanical and nano-tribological characterization of Ni-Co-BN nano-composite coating for bearing applications. *Tribol. Int.* **2023**, *180*, 108281. [CrossRef]
8. El-Shamy, O.A.A.; Deyab, M.A. Novel anticorrosive coatings based on nanocomposites of epoxy, chitosan, and silver. *Mater. Lett.* **2023**, *330*, 133298. [CrossRef]
9. Vikas, R.K.S.R.; Ramana, V.S.N.V.; Morampudi, P.; Bhavani, K.; Reddy, G.M. Unique approaches in developing novel nano-composites: Evaluating their mechanical and tribological characteristics. *J. Mech. Behav. Mater.* **2025**, *34*, 20250054. [CrossRef]
10. Tripathy, D.B.; Gupta, A. Nanocomposites as sustainable smart materials: A review. *J. Reinf. Plast. Compos.* 2024, *in press*. [CrossRef]
11. Protyai, M.I.H.; Saif, S.J.A.; Rahman, M.M.; Mural, H.; Bin Rashid, A. An overview of recent progress of green nano-composites for sustainable energy storage applications. *Eng. Rep.* **2026**, *8*, e70662. [CrossRef]
12. Antunes, M.; Arencón, D. Recent Developments in the Mechanical Behavior of Polymer-Based Composites. *Polymers* **2026**, *18*, 598. [CrossRef]
13. Chen, B.; Xu, M. Preparation and performance analysis of micro-nano composite coatings reinforced with WC particles. *Mater. Res. Express* **2024**, *11*, 026508. [CrossRef]
14. Liu, L.; Thangaraj, M.; Karmiris-Obratański, P.; Zhou, Y.; Annamalai, R.; Machnik, R.; Elsheikh, A.; Markopoulos, A.P. Optimization of Wire EDM Process Parameters on Coutting Inconel 718 Alloy with Zinc-Diffused Coating Brass Wire Electrode Using Taguchi-DEAR Technique. *Coatings* **2022**, *12*, 1612. [CrossRef]
15. Liu, Q.; Zhao, Y.; Fan, Z.; Yu, J.; Zeng, M. ZIF-67 Derived Co-Ni/C Composites for high-Efficiency Broadband Microwave Absorption. *Coatings* **2022**, *12*, 1667. [CrossRef]
16. Thangamani, G.; Thangaraj, M.; Anand, P.I.; Jayakumar, M.; Karkalos, N.E.; Papazoglou, E.L.; Karmiris-Obratański, P. Influence of Silver-Coated Tool Electrode on Electrochemical Micromachining Incoloy 825. *Coatings* **2023**, *13*, 963. [CrossRef]
17. Ashkinazi, E.E.; Fedorov, S.V.; Martyanov, A.K.; Sedov, V.S.; Popovich, A.F.; Bolshakov, A.P.; Sovyk, D.N.; Ryzhkov, S.G.; Khomich, A.A.; Zavedeev, E.V.; et al. Evolution of the Growth of a Micro-Nano Crystalline Diamond Film on an Axial Carbide Tool Model in Microwave Plasma. *Coatings* **2023**, *13*, 1156. [CrossRef]
18. Kiiamov, A.G.; Ivanova, A.G.; Solodov, A.N.; Cherosov, M.A.; Tayurskii, D.A.; Khannanov, A. Determination of the Size of Complex Iron Oxide Nanoparticles Using Various Physical Experimental Methods. *Coatings* **2023**, *13*, 1589. [CrossRef]
19. Korzekwa, J.; Niedźwiedź, M.; Dercz, G.; Cwynar, K.; Sowa, M.; Bara, M.; Simka, W. Methods of Distributing the IF-WS2 Modifier for Its Introduction into the Structure of the Al2O3 Aluminum Oxide Coating. *Coatings* **2024**, *14*, 883. [CrossRef]
20. Ibarra, J.J.; González, M.A.; Rodríguez, E.; Vásquez, G.I.; Medina, A.; Bernal, J.; Aguilar, C.; Velez, E.E. Enhanced Fracture Toughness of WC-CoCr Thermally Sprayed Coatings by the Addition of NiCrFeSiBC and Mo and Its Influence on Sliding Wear Behavior. *Coatings* **2024**, *14*, 1207. [CrossRef]
21. Tuhta, S.; Günday, F. Investigation of Steel Chimney Retrofitted with Nanocoating under Earthquake Excitation Using FEM. *Coatings* **2025**, *15*, 619. [CrossRef]
22. Yuce, I.; Tuhta, S. Behavior of Flexible Biogas Digester Made of PVC-Coated PET Polyester Fabric with Increased Durability by Nanocoating Under Northridge (1994) Earthquake Effect. *Coatings* **2025**, *15*, 1401. [CrossRef]

https://doi.org/10.3390/coatings16050548

coatings

MDPI

Article

Behavior of Flexible Biogas Digester Made of PVC-Coated PET Polyester Fabric with Increased Durability by Nanocoating Under Northridge (1994) Earthquake Effect

Ismail Yuce [1,*] and Sertaç Tuhta [2]

[1] Textile, Clothing, Shoe and Leather Department, Technical Science Vocational School, Trakya University, 22030 Edirne, Turkey

[2] Department of Civil Engineering, Faculty of Engineering, Ondokuz Mayis University, 55139 Samsun, Turkey; stuhta@omu.edu.tr

* Correspondence: ismailyuce@trakya.edu.tr; Tel.: +90-5063431481

Abstract: The purpose of this study is to ascertain how the dynamic reactions of a flexible biogas digester composed of PVC coated on PET polyester fabric are affected by nanocoating retrofitting. In this study, the effects of seismic movements on a ZrO_2-coated and uncoated PVC-PET biogas digester were investigated. A semi-sphere biogas digester that was 2.5 m high was chosen for development with nanocoating. Frequency values and mode shapes were acquired by first creating a three-dimensional finite element model of the biogas digester and then performing modal analysis in SAP2000 software. Later, the ground motion data from the 1994 Northridge earthquake were used to do linear assessments of the methane digester. Dynamic evaluations were then performed after a 0.25 mm ZrO_2 nanocoating was installed onto the digester's exterior. Finally, the dynamic behavior of the biogas digester, characterized by minimum and maximum von Mises stresses and corresponding displacements, was evaluated before and after the ZrO_2 coating. By the end of the study, the maximum von Mises stress had decreased from 60.47 kPa to 15.18 kPa, while the minimum von Mises stress at the digester bottom was reduced from 11.41 kPa to 4.47 kPa. Additionally, the displacements along the digester height, particularly the maximum values, were reduced from 0.057 mm to 0.013 mm. The reduction in frequency values demonstrated that ZrO_2 nanocoating improved the biogas digester's rigidity. This study makes it clear that nanocoatings will have an impact on dynamic parameters and be highly beneficial for improving biogas digesters.

Keywords: nanocoating; digester; FEM; durability; dynamic parameters

1. Introduction

Biogas digesters are sustainable energy production systems widely used in rural and industrial areas for renewable energy generation and waste management [1,2]. These digesters provide an economical solution by converting organic waste into biogas and organic fertilizer, thereby contributing to both energy efficiency and environmental protection [3].

Flexible biogas digesters, produced from polymer-based composite fabrics such as poly(ethylene terephthalate) (PET)/poly(vinyl chloride) (PVC) textiles, are preferred due to their ease of installation, low production cost, portability, and adaptability to different field conditions [4,5]. However, their flexible and thin-walled structure makes them vulnerable to environmental impacts such as temperature changes, ultraviolet radiation, chemical effects, and, especially, dynamic loads such as wind and earthquakes [4,6]. The structural safety of flexible biogas digesters has become an important research subject in recent years, particularly in earthquake-prone regions [7,8]. Unlike rigid reinforced concrete or steel structures, the dynamic behavior of polymer-based flexible digesters is highly nonlinear and their response depends on parameters such as membrane stiffness, geometric configuration, and material degradation over time [9].

Polyester fiber is a class of polymers containing repeating ester groups in the polymer chain. Polyester fiber sold commercially on the market is poly(ethylene terephthalate) [10]. PVC has a wide range of industrial applications. PVC is a highly popular polymer due to its low cost, excellent physical properties, compatibility with additives, wide range of applications, and versatility in processing techniques. PVC is one of the four main polymer classes commonly used in the coating of textile products. It is also widely used in the coating of polyester fabrics. While this coating provides impermeability and resistance against chemical factors, it is susceptible to aging, UV degradation, micro-cracking, and puncture under cyclic or sudden loading [11]. Therefore, strengthening methods are needed to extend the service life of such digesters and to ensure their safe performance under seismic loading. Similarly to the repair and retrofitting practices applied in civil engineering structures (chimneys, tanks, silos, bridges, etc.), methods such as fiber-reinforced polymer (FRP) wrapping, composite layer reinforcement, or ring-type strengthening have been reported in the literature [12–15]. In Obileke's [16] study, it was reported that biogas digesters made from HDPE experienced whole formation and material degradation over time because of UV radiation. This situation raises questions about the long-term practicality and cost-effectiveness of conventional reinforcement methods (e.g., FRP wrapping, composite layer addition) that can be applied to thin and flexible membrane-type digesters. In this context, nanotechnology-based coatings (nanocoatings) have emerged as an innovative approach. With the rapid developments in nanoscience in the last 25 years, nanomaterials are increasingly being integrated into structural elements to improve durability, strength, and functional performance [17–20].

Nanotechnology refers to the production and manipulation of materials at dimensions between 1 and 100 nanometers, where quantum effects, surface-to-volume ratios, and electrostatic interactions dominate. At these scales, materials exhibit unique mechanical, thermal, and chemical properties compared to their macro-scale forms [21]. Nanocoatings applied to structural materials can enhance resistance against abrasion, chemical corrosion, ultraviolet degradation, and water permeability, while also improving strength and toughness [22–24]. Magnesium oxide (MgO), zirconium oxide (ZrO_2), and carbon nanotubes (CNTs) are among the most widely studied nanomaterials in civil engineering. ZrO_2 nanomaterials stand out for their thermal resistance and durability, while CNTs exhibit high tensile strength and effective crack bridging properties [25–28]. In a study conducted by Dhineshbabu and colleagues, SiO_2 and ZrO_2/SiO_2 nanoparticles were obtained using the sol–gel method and coated onto cotton fabrics. It was observed that the flame retardancy of ZrO_2/SiO_2-coated fabrics was superior to that of SiO_2-coated fabrics. Furthermore, ZrO_2/SiO_2-coated fabrics exhibited excellent UV protection compared to SiO_2-coated fabrics. As a result of the study, it can be said that ZrO_2 particles enhance UV protection and flame retardancy [29]. In another study conducted by Jia and colleagues, the mechanical properties of ZrO_2-coated, basalt fabric-reinforced concrete were investigated. It was observed that ZrO_2-coated basalt fabric has a high strength retention ratio in long-term alkaline environments, and as a result the ZrO_2-coated basalt fabric reinforcement was observed to significantly enhance the mechanical properties of concrete [30].

Applications of nanotechnology in civil engineering have traditionally focused on improving concrete, steel, and composite materials, as seen in studies on high-strength concrete, corrosion-resistant steel, and FRP composites [31–35]. However, research on the use of nanocoatings in flexible, polymer-based structures such as biogas digesters is very limited. This reveals a significant research gap in the literature. Considering the increasing importance of renewable energy and sustainable structural solutions, extending the service life of flexible digesters through nanocoating reinforcement is both scientifically and practically significant.

In this study, the dynamic behavior of a flexible biogas digester made of PVC-coated PET fabric reinforced with a thin ZrO_2 nanocoating layer was investigated under seismic loading conditions. To achieve this, a 3D finite element model (FEM) of the digester was constructed using the SAP2000 (v26.2.0) software package. The digester geometry was defined as a cylindrical flexible structure with a dome-shaped roof, commonly used in field applications. The PVC-coated fabric material (0.003 m) was modeled with nonlinear membrane elements, while the nanocoating reinforcement was applied as a thin surface layer (0.00025 m) covering the outer face of the digester membrane. The model was validated using previous numerical and experimental studies on membrane structures.

The 1994 Northridge earthquake severely affected industrial facilities in the United States [36]. In the study, the seismic movement records of this earthquake were simulated. Dynamic analyses were performed both for the unreinforced digester and for the digester reinforced with ZrO_2 nanocoating. Comparative results were obtained in terms of maximum displacements, von Mises stress distributions, natural frequencies, and mode shapes. In addition, the effects of nanocoating on membrane stiffness and energy dissipation capacity were discussed.

The findings of this study are expected to contribute to filling the gap in the literature regarding the nanotechnology-based improvement of flexible polymeric biogas digesters, and to guide practical applications in earthquake-prone regions. With the integration of nanocoating technologies, it is aimed to increase the service life, durability, and seismic safety of such renewable energy systems.

This research is expected to contribute to the literature in the following three important ways:

1. To present one of the first comprehensive studies on flexible polymeric biogas digesters developed using nanocoating, thereby filling a gap in the literature. At the same time, filling this gap also highlights the innovative aspect of the study.
2. Demonstrating the effectiveness of ZrO_2 nanocoatings in enhancing structural performance, durability, and seismic safety.
3. Guiding future applications for sustainable and earthquake-resilient renewable energy systems in rural and industrial contexts.

Integrating nanotechnology into the improvement of flexible digesters, this study aims to improve their service life, mechanical reliability, and overall resilience, thereby supporting broader global goals in sustainable infrastructure and clean energy generation.

2. Description of the Flexible Biogas Digester

A flexible biogas digester with a height of 2.5 m is simulated for dynamic analysis. The digester wall thickness was designed as 0.003 m and its outer diameter was 5 m. This thickness was chosen because the thickness of commercially available membranes is 0.003 m [37]. Figure 1 gives the general structural features of the digester.

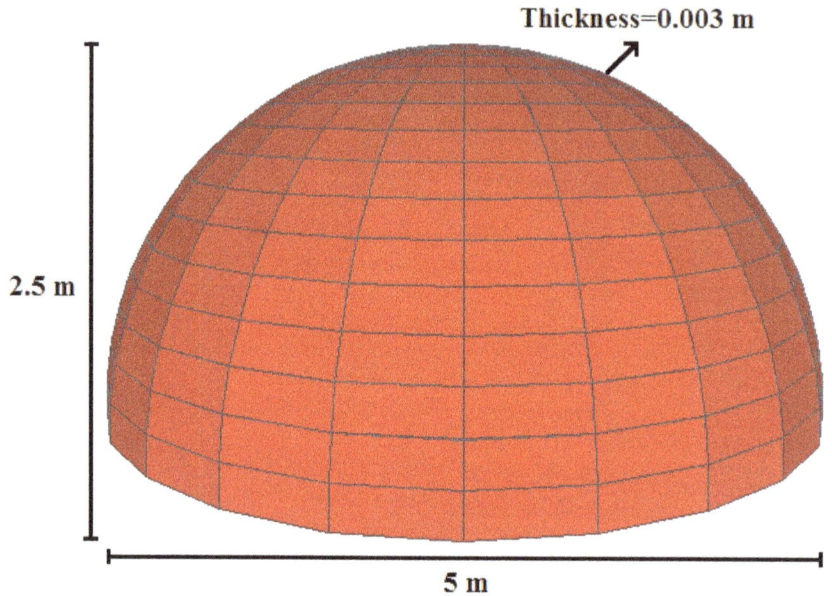

Figure 1. General properties of biogas digester.

3. Finite Element Modeling of the Digester

SAP 2000 software is used for the finite element analysis of linear and nonlinear three-dimensional structures [38]. In this study, this software was used to design three-dimensional finite element models of ZrO_2-coated and uncoated digesters. It can model different types of elements such as shell, solid, and frame components. Also, it includes advanced analysis tools for evaluating modal behavior, time–history responses, and response spectrum analyses. The software enables the definition of layered shell elements with distinct material properties, making it particularly suitable for simulating composite or coated structural systems such as polymer-based biogas digesters [39].

With the accurate determination of the mesh size, analytical modeling gives reliable results. Convergence analyses were performed to define the optimal mesh size of the flexible biogas digester. The three-dimensional finite element analysis of the analyzed digester was confirmed according to the literature data. The four-node shell element was used to model the digester. Each shell element is composed of nine nodes, and each node has six degrees of freedom—three translational (along the x, y, and z axes) and three rotational (about the same axes) (Figure 2).

The geometry and node arrangement of this element type are presented in Figure 2, where a layered shell element was utilized to simulate the ZrO_2-coated structure. This element type can accommodate multiple material layers, each characterized by unique orientations and orthotropic mechanical properties. This element type is especially well-suited for modeling layered thick shells or solid structures. Figure 2 presents a schematic representation of the shell element.

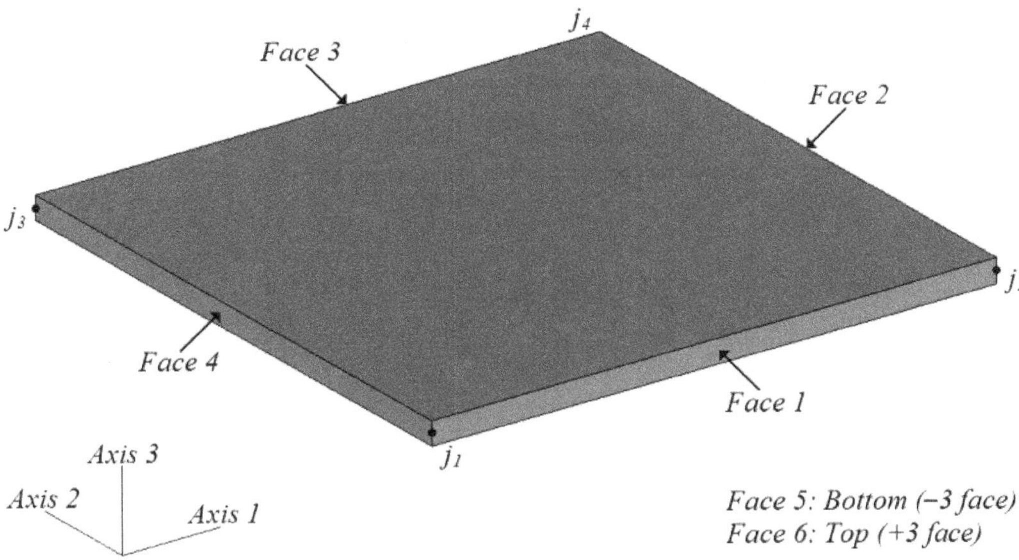

Figure 2. Diagram and characteristics of the shell element (SAP2000, 1997) [38].

Reliable analytical modeling of tall, slender structures like biogas digesters requires accurate characterization of material properties and boundary conditions. In this present study the materials were assumed to exhibit linear elastic behavior, with the corresponding properties of fabric and ZrO_2 defined accordingly. SAP2000 modeling of the biogas digester, fabricated from PVC-coated PET polyester fabric, required specification of the modulus of elasticity, Poisson's ratio, and unit weight. The elastic modulus, Poisson's ratio, and mass per unit volume of fabric are presented in Table 1.

Table 1. Material properties used in analysis.

Material	Modulus of Elasticity (kN/m^2)	Poisson's Ratio	Mass Per Unit Volume (kN/m^3)
Fabric *	3.5×10^6	0.40	1.4
ZrO_2	1.75×10^8	0.27	5.6

* PVC Coated PET Polyester Fabric.

Nanomaterials, such as ZrO_2, have been demonstrated to enhance the mechanical performance of reinforced concrete, masonry, steel, and timber structures, particularly under flexural and shear stresses. These enhancements are valuable in retrofitting applications involving increased loading capacity, functional modifications, defect remediation, seismic strengthening, and compliance with updated structural standards. A ZrO_2 nanocoating was applied to improve the selected biogas digester, with material characteristics summarized in Table 1.

An initial meshing scheme was applied to the model. However, this produced excessively large linear transient analyses and model volumes that hindered effective visualization. Consequently, the mesh size was refined to achieve a balance between computational efficiency and accuracy. The final finite element model comprised 384 shell elements and 385 nodes, which provided a manageable yet sufficiently detailed representation for subsequent analyses.

The all-digester wall was coated with 0.00025 m-thick ZrO_2 (single layer). The thickness of the ZrO_2 coating (0.00025 m) was thin compared to the thickness of the digester wall (0.003 m). When the literature is examined, it can be said that a coating thickness of 0.00025 m (between 0.0002 and 0.0005 m) is sufficient to strengthen the digester [40]. Attention was paid to tensile and compressive rigid behaviors of the elements. At the same time, the single direction of the fibers (zero degree) is taken as a basis. It is assumed that all degrees of freedom at the base of the digester are connected to the ground and the digester is fixed to the ground. Perfect bonding between the ZrO_2-layered shell elements and fabric elements is achieved by interconnecting adjacent nodes. This ensures that the shell elements (PET/PVC fabric and ZrO_2) share common nodal points. Additionally, the ZrO_2-layered shell elements and fabric shell elements were meshed at identical sizes to achieve perfect interaction. Figure 3 shows images of ZrO_2-reinforced and non-reinforced biogas digesters.

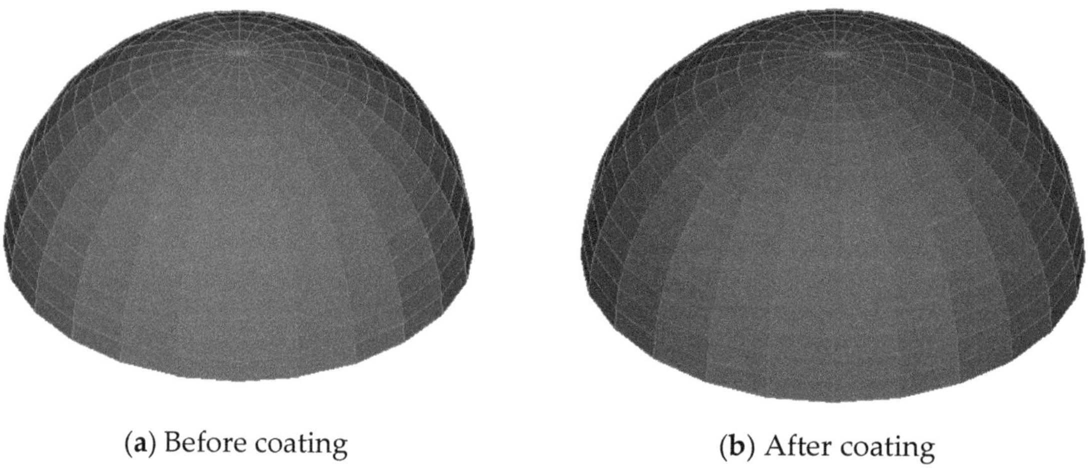

Figure 3. Finite element model of biogas digester before and after ZrO_2 coating.

Generally, industrial biogas digesters exhibit thin and homogeneous mass distributions. Therefore, higher modes significantly affect these digesters' dynamic behaviors. All modes with cumulative modal mass participation of at least 90% were considered in the time–history analyses. Modal analyses were carried out to determine the frequencies and mode shapes of the digester.

A total of five frequencies of digesters before and after ZrO_2 coating were obtained in the range of 54.82–68.25 Hz and 112.35–146.41 Hz, respectively. In Figure 4, the first five mode shapes and frequencies of the digesters obtained from modal analyses were illustrated. From Figure 4, it can be seen that the same mode shapes were obtained in the modal analyses before and after ZrO_2 coating. Although an increase in frequency generally means a decrease in displacement demand, the jump observed in the values here suggests a significant increase in global stiffness.

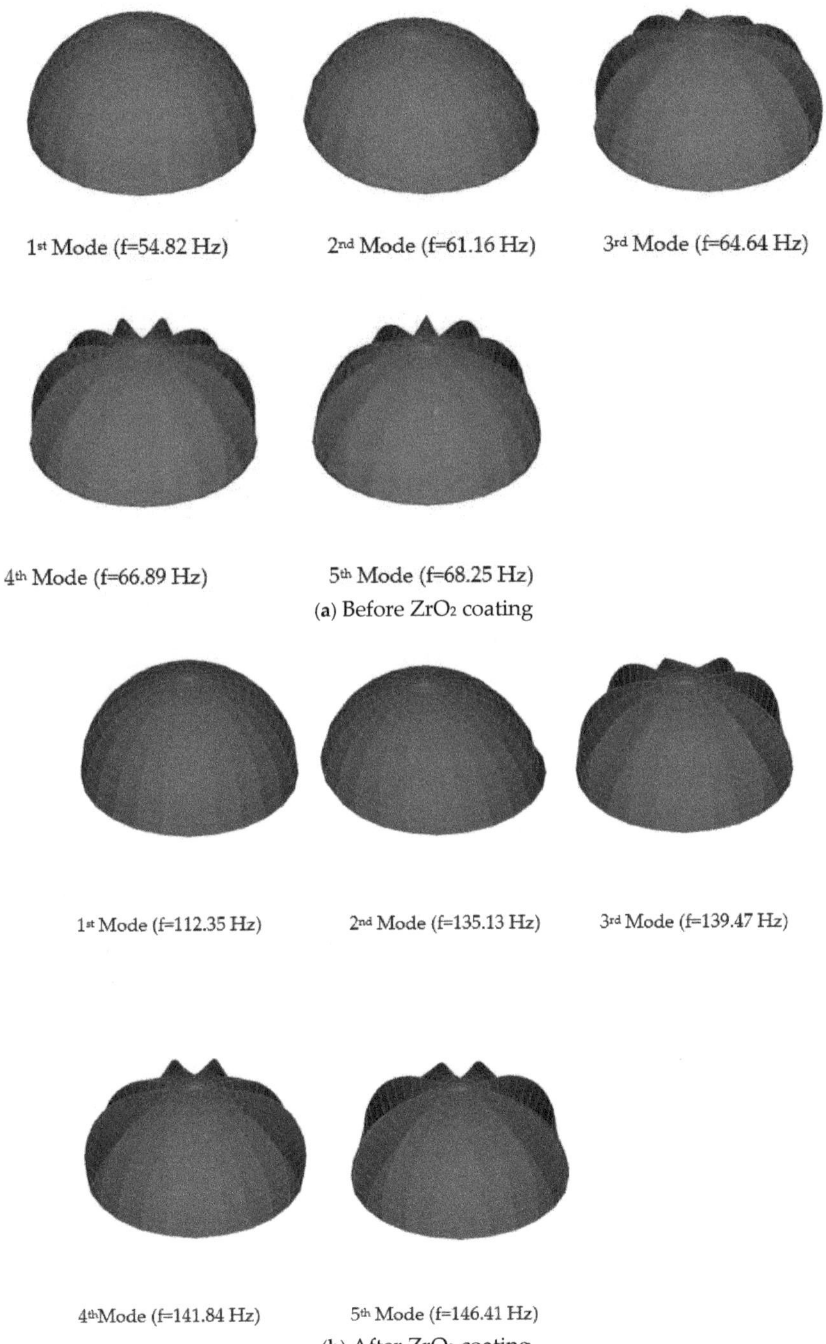

Figure 4. Mode shapes and frequencies of biogas digesters before (green) and after ZrO_2 (gray) coating.

4. Dynamic Analyses of the Biogas Digester

Dynamic analyses were conducted to evaluate the seismic performance of the flexible biogas digester both before and after the application of the ZrO_2 nanocoating. The linear time–history analyses used the north–south ground motion records of the 1994 Northridge earthquake, characterized by a peak ground acceleration (PGA) of 0.842 g (Figure 5). These records are widely recognized for their strong near-field characteristics, making them suitable for assessing the behavior of light, flexible membrane structures such as polymer-based digesters.

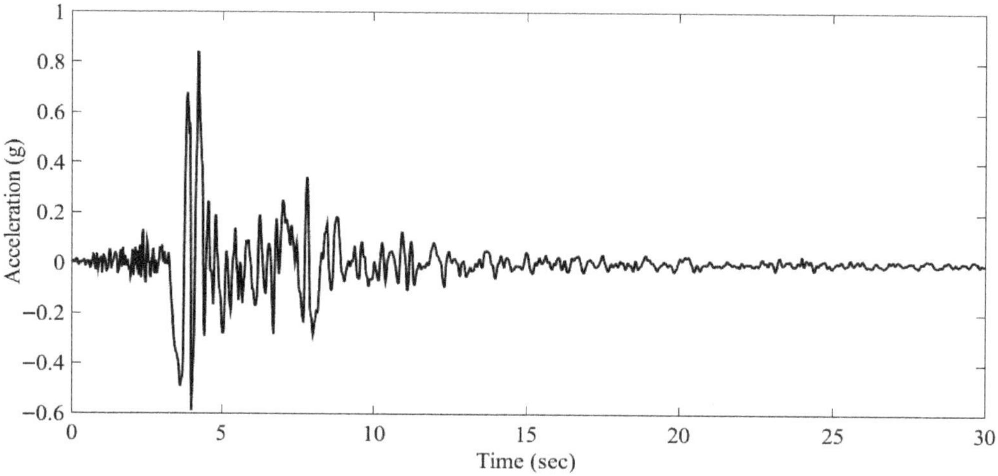

Figure 5. Time history of ground motion acceleration from the 1994 Northridge earthquake (PGA = 0.842 g).

A Rayleigh damping model with 5% critical damping was adopted, calibrated between the first and fifth vibration modes. This approach ensured that the model realistically captured the energy-dissipating behavior typical of thin polymeric membranes, which exhibit limited inherent damping compared to reinforced concrete or steel structures.

4.1. Displacement Response

Figure 6 presents the maximum horizontal displacement contours along the digester height. Before applying the ZrO_2 coating, the digester exhibited a maximum top displacement of 0.057 mm. After coating, this value decreased significantly to 0.013 mm, corresponding to a ~77% reduction. The inclusion of a ZrO_2 nanolayer increases the composite modulus of the membrane system, leading to higher resistance against lateral deformation. Even though the coating's thickness (0.00025 m) is small relative to the membrane thickness (0.003 m), nanomaterials can meaningfully alter stiffness due to their superior elastic properties [41].

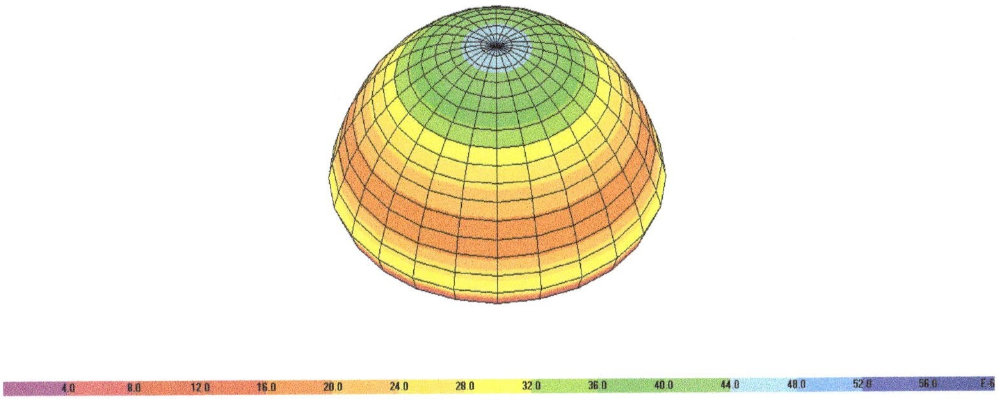

(**a**) Before ZrO_2 coating

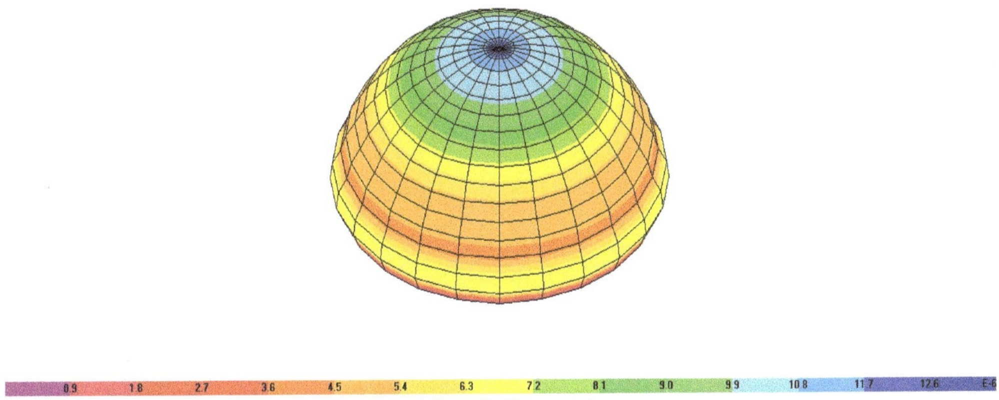

(**b**) After ZrO_2 coating

Figure 6. Comparison of maximum displacement contours before and after ZrO_2 coating.

The displacement attenuation observed in the coated digester indicates a more stable dynamic behavior, with reduced susceptibility to buckling-type deformations that can compromise long-term serviceability in real field conditions.

4.2. Stress Distribution and Critical Zones

Von Mises stress contours (Figure 7) reveal the regions most affected by seismic loading. Before reinforcement, the digester exhibited a maximum von Mises stress of 60.47 kPa at the dome–cylinder transition region, where curvature changes typically induce stress concentrations. The minimum stress at the base region was 11.41 kPa. After coating, the maximum and minimum stresses dropped to 15.18 kPa and 4.47 kPa, respectively. This represents a 74.9% reduction in maximum stress and a 60.8% reduction in minimum stress. The reduction in stress levels demonstrates the improved load-bearing capacity of the coated structure.

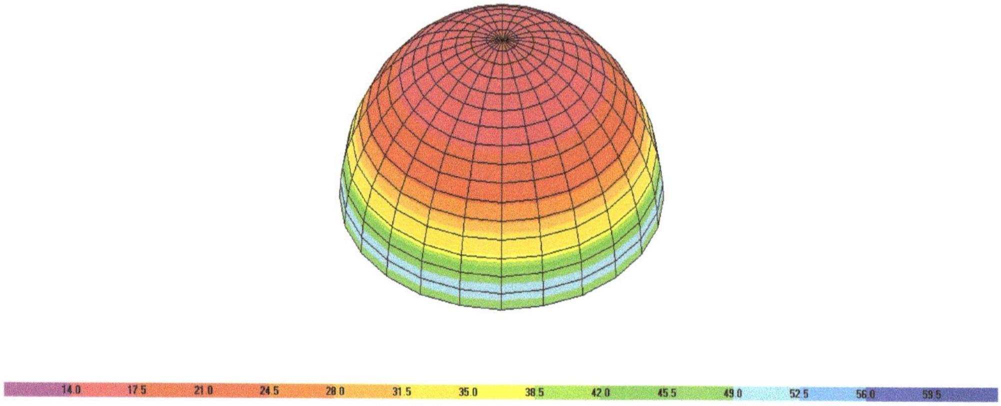

(**a**) Before ZrO_2 coating

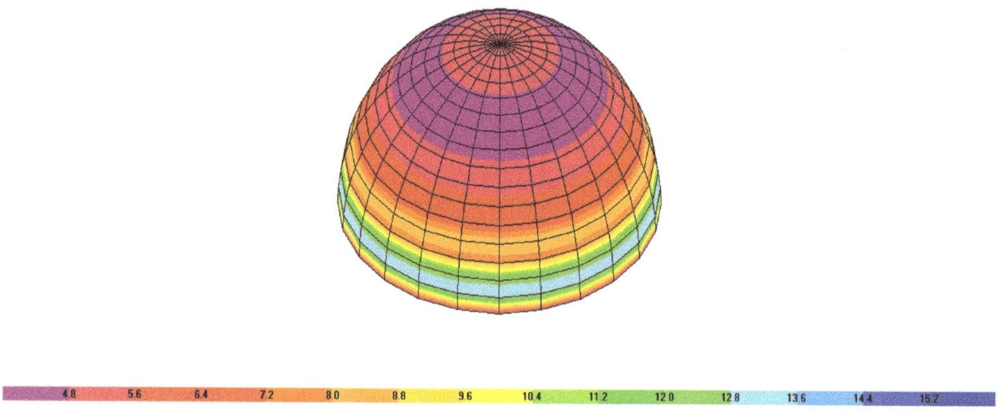

(**b**) After ZrO_2 coating

Figure 7. Von Mises stress contours (minimum–maximum) of the biogas digester before and after ZrO_2-coating application.

The results collectively demonstrate that ZrO_2 nanocoating provides substantial performance improvements. These improvements can be summarized as enhanced service life, reduced maintenance requirements, and improved seismic resilience.

5. Conclusions

This study investigated the dynamic behavior of fabric biogas digesters reinforced with ZrO_2 coating using finite element analysis in SAP2000 and the 1994 Northridge earthquake data. The key findings are:

I. **Enhanced structural rigidity**: ZrO_2 coating increased natural frequencies from 54.82–68.25 Hz to 112.35–146.41 Hz, indicating improved stiffness without significant mass addition.

II. **Reduced displacement**: Maximum horizontal displacement at the digester's top decreased by 77% (from 0.057 mm to 0.013 mm), demonstrating superior performance compared to conventional reinforcement methods.

III. **Improved stress distribution**: Von Mises stress at critical bottom regions reduced significantly—minimum stress decreased from 11.41 kPa to 4.47 kPa and maximum stress from 60.47 kPa to 15.18 kPa—minimizing stress concentration at ground interface zones.
IV. **Practical benefits**: Increased rigidity reduces crack formation, corrosion, and maintenance costs while maintaining a favorable strength-to-weight ratio.
V. **Broader implications**: Nanocoatings offer superior corrosion resistance, easy applicability, extended service life, and fire resistance, positioning them as viable alternatives to traditional retrofitting techniques for various biogas digester applications.

This pioneering study demonstrates that ZrO_2 nanocoating significantly enhances the seismic performance of fabric biogas digesters. Further research is needed to expand the literature on nanocomposite applications in biogas infrastructure.

Author Contributions: Methodology, I.Y.; Software, S.T.; Investigation, I.Y. and S.T.; Data curation, S.T.; Writing—original draft, I.Y. and S.T.; Writing—review & editing, I.Y. and S.T.; Funding acquisition, I.Y. and S.T. All authors have read and agreed to the published version of the manuscript.

Funding: This research received no external funding.

Institutional Review Board Statement: Not applicable.

Informed Consent Statement: Not applicable.

Data Availability Statement: Data are contained within the article.

Conflicts of Interest: The authors declare no conflict of interest.

Abbreviations

The following abbreviations are used in this manuscript:

PVC	Poly(vinyl chloride)
PET	Poly(ethylene terephthalate)
FEM	Finite Element Method
FRP	Fiber Reinforced Polymer
UV	Ultraviolet
HDPE	High Density Polyethylene
PGA	Peak Ground Acceleration
CNTs	Carbon Nanotubes

References

1. Martins das Neves, L.C.; Converti, A.; Vessoni Penna, T.C. Biogas production: New trends for alternative energy sources in rural and urban zones. *Chem. Eng. Technol. Ind. Chem. Plant Equip. Process Eng. Biotechnol.* **2009**, *32*, 1147–1153. [CrossRef]
2. Abanades, S.; Abbaspour, H.; Ahmadi, A.; Das, B.; Ehyaei, M.A.; Esmaeilion, F.; Assad, M.E.H.; Hajilounezhad, T.; Jamali, D.H.; Hmida, A.; et al. A critical review of biogas production and usage with legislations framework across the globe. *Int. J. Environ. Sci. Technol.* **2022**, *19*, 3377–3400. [CrossRef] [PubMed]
3. Drosg, B.; Fuchs, W.; Al Seadi, T.; Madsen, M.; Linke, B. *Nutrient Recovery by Biogas Digestate Processing*; IEA Bioenergy: Dublin, Ireland, 2015; Volume 2015, p. 711.
4. Obileke, K.; Onyeaka, H.; Nwokolo, N. Materials for the design and construction of household biogas digesters for biogas production: A review. *Int. J. Energy Res.* **2021**, *45*, 3761–3779. [CrossRef]
5. Ghiandelli, M. Development and Implementation of Small-Scale Biogas Balloon Biodigester in Bali, Indonesia. Master's Thesis, KTH School of Industrial Engineering and Management, Stockholm, Sweden, 2017.
6. Asadi, H.; Uhlemann, J.; Stranghoener, N.; Ulbricht, M. Tensile strength deterioration of PVC coated PET woven fabrics under single and multiplied artificial weathering impacts and cyclic loading. *Constr. Build. Mater.* **2022**, *342*, 127843. [CrossRef]

7. Cheng, S.; Li, Z.; Mang, H.P.; Huba, E.M.; Gao, R.; Wang, X. Development and application of prefabricated biogas digesters in developing countries. *Renew. Sustain. Energy Rev.* **2014**, *34*, 387–400. [CrossRef]
8. Budiman, I. The role of fixed-dome and floating drum biogas digester for energy security in Indonesia. *Indones. J. Energy* **2020**, *3*, 83–93. [CrossRef]
9. Chandraul, A.; Murari, V.; Kumar, S. A review on dynamic analysis of membrane based space structures. *Adv. Space Res.* **2024**, *74*, 740–763. [CrossRef]
10. Polyester. Available online: https://en.wikipedia.org/wiki/Polyester (accessed on 4 November 2025).
11. Sen, A.K. *Coated Textiles: Principles and Applications*, 2nd ed.; CRC Press, Taylor & Francis Group: Boca Raton, FL, USA, 2008.
12. Meier, U. Strengthening of Structures Using Carbon Fibre/Epoxy Composites. *Constr. Build. Mater.* **1995**, *9*, 341–351. [CrossRef]
13. Mirmiran, A.; Zagers, K.; Yuan, W. Nonlinear Finite Element Modeling of Concrete Confined by Fiber Composites. *Finite Elem. Anal. Des.* **2000**, *35*, 79–96. [CrossRef]
14. Van Den Einde, L.; Zhao, L.; Seible, F. Use of FRP Composites in Civil Structural Applications. *Constr. Build. Mater.* **2003**, *17*, 389–403. [CrossRef]
15. Chen, C.H. Seismic Assessment of a Retrofitted Chimney by FEM Analysis and Field Testing. *Int. J. Struct. Stab. Dyn.* **2004**, *4*, 337–359. [CrossRef]
16. Obileke, K.; Mamphweli, S.; Meyer, E.L.; Makaka, G.; Nwokolo, N. Design and fabrication of a plastic biogas digester for the production of biogas from cow dung. *J. Eng.* **2020**, *2020*, 1848714. [CrossRef]
17. Yuce, I.; Canoglu, S.; Yukseloglu, S.M.; Li Voti, R.; Cesarini, G.; Sibilia, C.; Larciprete, M.C. Titanium and silicon dioxide-coated fabrics for management and tuning of infrared radiation. *Sensors* **2022**, *22*, 3918. [CrossRef]
18. Farooq, S.; Akhtar, A.; Faisal, S.; Husain, M.D.; Siddiqui, M.O.R. Durable multifunction finishing on polyester knitted fabric by applying zinc oxide nanoparticles. *Pigment Resin Technol.* **2025**, *54*, 1006–1015. [CrossRef]
19. Sharouf, H.M.; Tuhmaz, G.; Saffour, Z. Preparation of Zinc Oxide Nano-Layer on Cotton Fabric for UV Protection Application. *Al-Nahrain J. Eng. Sci.* **2019**, *22*, 18–21. [CrossRef]
20. Mankodi, D.H.; Agarwal, D.B. Studies on nano–UV protective finish on apparel fabrics for health protection. *Res. J. Text. Appar.* **2011**, *15*, 11–20. [CrossRef]
21. Szczyglewska, P.; Feliczak-Guzik, A.; Nowak, I. Nanotechnology–general aspects: A chemical reduction approach to the synthesis of nanoparticles. *Molecules* **2023**, *28*, 4932. [CrossRef]
22. Farooq, S.A.; Raina, A.; Mohan, S.; Arvind Singh, R.; Jayalakshmi, S.; Irfan Ul Haq, M. Nanostructured coatings: Review on processing techniques, corrosion behaviour and tribological performance. *Nanomaterials* **2022**, *12*, 1323. [CrossRef]
23. Gu, Y.; Xia, K.; Wu, D.; Mou, J.; Zheng, S. Technical characteristics and wear-resistant mechanism of nano coatings: A review. *Coatings* **2020**, *10*, 233. [CrossRef]
24. Pourhashem, S.; Saba, F.; Duan, J.; Rashidi, A.; Guan, F.; Nezhad, E.G.; Hou, B. Polymer/Inorganic nanocomposite coatings with superior corrosion protection performance: A review. *J. Ind. Eng. Chem.* **2020**, *88*, 29–57. [CrossRef]
25. Meena, K.L.; Vidyasagar, C.S.; Benny Karunakar, D. Mechanical and tribological properties of MgO/multiwalled carbon nanotube-reinforced zirconia-toughened alumina composites developed through spark plasma sintering and microwave sintering. *J. Mater. Eng. Perform.* **2022**, *31*, 682–696. [CrossRef]
26. Gómez-Rodríguez, C.; Fernández-González, D.; García-Quiñonez, L.V.; Castillo-Rodríguez, G.A.; Aguilar-Martínez, J.A.; Verdeja, L.F. MgO refractory doped with ZrO_2 nanoparticles: Influence of cold isostatic and uniaxial pressing and sintering temperature in the physical and chemical properties. *Metals* **2019**, *9*, 1297. [CrossRef]
27. Valdez-Salas, B.; Salvador-Carlos, J.; Beltrán-Partida, E.A.; Castillo-Sáenz, J.; Chairez-González, J.; Curiel-Álvarez, M. Zirconium Nanostructures Obtained from Anodic Synthesis By-Products and Their Potential Use in PVA-Based Coatings. *Ceramics* **2025**, *8*, 74. [CrossRef]
28. Bogas, J.A.; Ahmed, H.H.; Diniz, T. Influence of cracking on the durability of reinforced concrete with carbon nanotubes. *Appl. Sci.* **2021**, *11*, 1672. [CrossRef]
29. Dhineshbabu, N.R.; Manivasakan, P., Yuvakkumar, R.; Prabu, P.; Rajendran, V. Enhanced functional properties of ZrO_2/SiO_2 hybrid nanosol coated cotton fabrics. *J. Nanosci. Nanotechnol.* **2013**, *13*, 4017–4024. [CrossRef]
30. Jia, M.; Xie, W.; Yu, K.; Qian, K. Preparation and characterization of basalt fabrics coated with zirconia obtained by sol-gel method for textile reinforced concrete. *J. Text. Inst.* **2022**, *113*, 527–540. [CrossRef]
31. Matthys, S.; Toutanji, H.; Taerwe, L. Stress–Strain Behavior of Large-Scale Circular Columns Confined with FRP Composites. *J. Struct. Eng.* **2006**, *132*, 123–133. [CrossRef]
32. Zhao, X.L.; Zhang, L. State-of-the-Art Review on FRP Strengthened Steel Structures. *Eng. Struct.* **2007**, *29*, 1808–1823. [CrossRef]
33. Pallares, F.J.; Ivorra, S.; Pallares, L.; Adam, J.M. Seismic Assessment of a CFRP-Strengthened Masonry Chimney. *Proc. Inst. Civ. Eng. Struct. Build.* **2009**, *162*, 291–299. [CrossRef]

34. Silvestre, J.; Silvestre, N.; De Brito, J. Review on concrete nanotechnology. *Eur. J. Environ. Civ. Eng.* **2016**, *20*, 455–485. [CrossRef]
35. Shi, X. On the use of nanotechnology to manage steel corrosion. *Recent Pat. Eng.* **2010**, *4*, 44–50. [CrossRef]
36. Jonczyk, J. Lessons Learned from the Northridge Earthquake (January 1994) with Respect to Design and Code Regulation on the Example of Steel Frame Behaviour (No. NEA-CSNI-R--2002-22, pp. 15–23). Available online: https://www.oecd-nea.org/upload/docs/application/pdf/2020-01/csni-r2002-22.pdf (accessed on 4 November 2025).
37. A1 Sports VARIES PVC Coated Tarpaulin Fabric, Thickness: 3 mm. Available online: https://www.aajjo.com/product/a1-sports-varies-pvc-coated-tarpaulin-fabric-thickness-3-mm-in-mumbai-a1-sports-machines (accessed on 4 November 2025).
38. Tuhta, S.; Günday, F. Investigation of Steel Chimney Retrofitted with Nanocoating Under Earthquake Excitation Using FEM. *Coatings* **2025**, *15*, 619. [CrossRef]
39. Surpam, M.; Tanawade, A. Analysis and Design of Pre-engineered Building Structure Using SAP2000. In *Dynamic Behavior of Soft and Hard Materials*; Springer Proceedings in Materials, Volume 36; Springer: Singapore, 2024; Volume 3, p. 22. [CrossRef]
40. Duong, X.Q.; Cao, N.V.; Lee, W.S.; Park, M.Y.; Chung, J.D.; Bae, K.J.; Kwon, O.K. Effect of coating thickness, binder and cycle time in adsorption cooling applications. *Appl. Therm. Eng.* **2021**, *184*, 116265. [CrossRef]
41. Samad, U.A.; Alam, M.A.; Abdo, H.S.; Anis, A.; Al-Zahrani, S.M. Synergistic effect of nanoparticles: Enhanced mechanical and corrosion protection properties of epoxy coatings incorporated with SiO_2 and ZrO_2. *Polymers* **2023**, *15*, 3100. [CrossRef]

coatings

MDPI

Article

Investigation of Steel Chimney Retrofitted with Nanocoating Under Earthquake Excitation Using FEM

Sertaç Tuhta [1] and Furkan Günday [2,*]

[1] Department of Civil Engineering, Faculty of Engineering, Ondokuz Mayis University, 55139 Samsun, Turkey; stuhta@omu.edu.tr
[2] Department of Civil Engineering, Faculty of Engineering, Giresun University, 28200 Giresun, Turkey
* Correspondence: furkan.gunday@giresun.edu.tr

Abstract: The aim of this research is to determine the effect of nanocoating retrofitting on the dynamic responses of industrial steel chimneys. A 20 m high industrial steel chimney was selected for nanocoating retrofitting. First, a three-dimensional finite element model of the chimney was created, and modal analysis was performed in SAP2000 software, and frequency values and mode shapes were obtained. Later, linear analyses of the chimney were made using the 1995 Kobe earthquake ground motion recording. Then, the outer surface of the chimney was retrofitted with a 1 mm MgO nanocoating, and dynamic analyses were made. Finally, the chimney's dynamic responses before and after the MgO retrofit, such as minimum–maximum von Mises stresses and displacements, were compared. At the end of the study, it was observed that the displacements along the chimney height, especially maximum displacements, decreased from 150 mm to 140 mm, and the minimum von Mises stress at the bottom of the chimney decreased from 37.20 kPa to 30.19 kPa, and the maximum von Mises stress from 1519.27 kPa to 1332.33 kPa stresses in the analyses results. The decrease in frequency values revealed that MgO nanocoating had a positive effect on the stiffness of the chimney. It can be understood from this study that nanocoatings will be very useful in retrofitting industrial steel chimneys and have an effect on dynamic parameters.

Keywords: nanocoating; chimney; FEM; earthquake; dynamic parameters

1. Introduction

Industrial steel chimneys are tall and slender structures that need repair and strengthening in their service lives. Today, nanocoatings are becoming a popular material used for the retrofitting of steel structures. Steel chimneys are industrial structures that remove waste gases from factories. Their heights, geometries and dimensions vary according to the needs of the facilities. Industrial chimneys are made of masonry, steel, reinforced concrete and composite materials [1–3]. As the need for high chimneys increased, those produced from steel and its alloys came to the fore. With developing technology, very important progress has been made in the field of corrosion. The fact that they are more resistant to external forces such as earthquakes and wind also causes them to be frequently preferred [4]. As with other materials, steel chimneys need repair or strengthening throughout their service life to maintain adequate ductility over time [5]. Quite different methods are used for the repair and strengthening of structures of this type. Conventionally, one of the most frequently used methods are operations that address the entire chimney, as well as intervening in parts that require repair. The most used of these processes involves as intervening in the parts requiring repair by using FRP-based materials and if there is no structural obstacle, supporting the section with rings made of steel or composite materials [6–9]. In study [10], samples representing RC chimneys were produced and tested under increasing horizontal monotonic loads and evaluated in terms of carrying capacity, crack formation, collapse mechanism, curvature ductility, moment capacity and energy absorption capacity. The test results showed that there was an average increase of 11.5% in the samples without FRP and 27.6% in the samples with FRP wrapping compared to the theoretical moment capacity. However, sudden collapse occurred in the samples wrapped with FRP, which revealed the need for additional design and application developments in such applications. In reference [11], the Chi-Chi earthquake in Taiwan in 1999 caused serious structural damage to a 250 m high chimney located in the central part of the country. In this study, experimental and numerical analysis methods were used together to evaluate the dynamic behavior, collapse mechanism and seismic performance of the chimney after repair. The dynamic properties of the chimney were determined by ambient vibration tests and system identification methods; the instrumental variable subspace approach was applied in this process. In addition, the three-dimensional dynamic behavior of the chimney was modeled with the finite element method using SAP2000 V26 software. The buckling capacity of the structure was evaluated according to the ASCE 1975 standard [12], and the calculated stress values were found to be consistent with the damage observed in the steel flues. The obtained findings demonstrate the reliability of the modeling and provide important information about the seismic performance of the chimney. On the other hand, it is seen that the use of nanomaterials, which are the result of working with developing technology, gives very effective results. Similarly, the use of FRP-based materials is highly preferred for retrofitting [13–17]. According to the seismic performance level required in the chimney, increasing the cross-section, supporting it with ring-like elements from the inside or outside, adding dampers, using FRP and using nanomaterials are among the methods that give effective results in retrofitting [18–23]. Today, in addition to conventional applications, various new methods have emerged with the widespread use of nanotechnological materials in the field of civil engineering. While these methods are based on the direct use of nanomaterials, it has also become possible to apply structural elements produced from these materials. It is difficult to make a broad definition, as nanotechnology, the foundations of which were laid approximately 50 years ago, is developing very rapidly. However, according to the definition agreed upon in every branch of science, nanotechnology is the manufacture of materials and products by controlling matter with its most fundamental particle (nanocoatings, nanoparticles, etc.). At the atomic, molecular or nanoscale levels, it can be expressed as the use of very small particles of material to create new large-scale materials. Nanotechnology is defined as the study and application of structures ranging in size from 1 nanometer (nm) to 100 nanometers. During nano-level processing, various problems arise such as gravity becoming negligible, electrostatic forces and quantum effects. Another point to be noted is that the use of nano-scale particles causes the material properties to change as a result of the increase in the ratio of atoms on the surface to those inside. Nanotechnology can produce materials with many unique properties that can improve existing building materials, such as lighter and stronger structural composites, low-maintenance building elements, fabrication in complex geometric shapes, and products with better thermal insulation properties. The incorporation of nanomaterials into the composition of some materials such as concrete, steel and composites will benefit air conditioning and energy efficiency. In addition, nanomaterials applied to the surfaces of building structural elements contribute to the elimination of structural deficiencies. Thanks to nanotechnology, concrete with a high strength, steel that can be made more durable and easier to assemble and wooden elements with a long durability can be produced. Due to the new potential applications of nanometer-scale particles, nanomaterials and nanotechnologies have become topics of great scientific interest. The construction sector lags behind other major industrial sectors in terms of the potential to benefit from nanotechnology. In addition to the use of nanotechnology in construction materials, today's popular topics regarding the potential uses of nanotechnology in the construction and development of construction materials are adding nanoparticles to increase the strength and durability of cement composites, using carbon nanotubes and nanofibers, producing low-cost and corrosion-resistant steel and the like. The use of nanotechnological materials in buildings not only prolongs their service life but also provides extra protection against various factors such as energy saving, fire resistance and corrosion. There are studies showing that the performance of building elements produced using materials such as cement-based composites, concrete, mortar, etc., is increased in terms of strength with the addition of nanomaterials [24–28]. In recent years, developments in structural engineering have made it possible to build bold structures with record spans and heights. However, significant challenges are encountered in the construction of such structures, such as the limitations of traditional materials and the need for continuous monitoring under complex dynamic loads. Considering the responsibilities for human life and the budget of the structures, continuous monitoring is inevitable to ensure the safe and efficient operation of structures. In this context, in the field of structural engineering, smart structures and materials with self-monitoring properties stand out among the most promising solution proposals. Although it is an area that has not yet been fully explored, current research reveals that composites reinforced with carbon-based nanomaterials have great potential as structural materials with a self-sensing capacity. Nanomaterials significantly improve the microstructure, mechanical properties and load-related properties of structural materials, thus increasing the durability and performance of structures. Therefore, the use of nanomaterials in structural engineering has become an important research topic in areas such as mechanical and electrical properties and durability [29]. Current nanoscience and nanotechnology researchers are examining these new features in depth. Because macro properties can change at the nanoscale, a significant number of new materials and processes can be created. Discussions regarding the application of nanotechnology in civil engineering, especially in construction, are critical. However, many advances in nanotechnology have the potential to be applied in the field of civil engineering. By using nanotechnological materials, various negative effects (via self-healing and waterproof concrete, self-cleaning plaster, etc.) that occur over time on building elements can be prevented. According to the studies reviewed, carbon nanotubes, one of the most widely used nanomaterials in civil engineering, are cylindrical carbon tubes that take their name from their nanometer diameter. They can be several millimeters long and have one "layer" or wall (single-walled nanotubes) or multiple walls (multi-walled nanotubes). Nanotubes have exceptional strength, unique electrical properties and efficient thermal properties. They are extremely popular because of their properties. Magnesium oxide nanoparticles, another nanomaterial especially for heat resistance, are also used to improve the properties of building elements. Most of the literature studies are on the reinforcement of reinforced concrete bridges, water tanks and chimneys with FRP-based composites. Studies on steel structures have generally focused on silos and water tanks. It is known that there are a limited number of studies on wood and other material types. It has been observed that studies in the field of civil engineering have emerged in the last ten years on nanotechnology. The majority of these studies are on the incorporation of nanoparticles into material structures and its consequences. Studies directly on the use of nanocoatings in civil engineering structures are limited [30]. It has been determined that there are reasons, such as the development of technology and the lack of application possibilities, for the small number of studies. In article [31], the dynamic behavior of glass fiber-reinforced epoxy nanocomposites was investigated. In the study, the effects of nanoclay and nanosilica additives on the mechanical properties and natural frequencies of epoxy composites were analyzed. It was observed that nanosilica and nanoclay additives added at different rates improved the dynamic properties of the composites and increased the mechanical performance of the material. In particular, it was found that 1% nanosilica additive increased the natural frequencies, but 3% nanoclay additive did not provide a significant improvement. These findings show that glass fiber-reinforced epoxy nanocomposites can be used as an effective reinforcement material to become more durable under dynamic loads and vibrations. The study aims to increase the potential of using such materials in future engineering applications by determining the effect of nanomaterials on epoxy-based composites [31]. In study [32], it was investigated how the 3D nanofiber network structure formed using bacterial cellulose and protonated chitosan improved the mechanical and thermal properties of natural fiber nanocomposites. The results showed that this structure increased the durability and flexibility of the composites [32].

This study provides an economic evaluation of the use of nanomaterials in concrete structures in the construction industry. The effective material properties of nanocomposite concrete were determined using the Halpin–Tsai micromechanical theory, and the effects of nanoparticles on mechanical properties such as elastic modulus, compressive strength and reinforcement percentage were investigated. The role of nanoparticles in the overall cost structure of concrete was analyzed using Euler theory and Hamilton's principle. The results showed that increasing the volume percentage of nanoparticles up to 1% increased the elastic modulus by 78%, while reducing the reinforcement ratio by 142%. In economic terms, it was found that the increase in the cost of concrete by adding nanoparticles was approximately USD 1 per ton, but this cost increase was quite low compared to the mechanical improvements provided [33]. MgO coating can be applied by different methods. These methods are generally thermal spray (plasma or flame spray) application, slurry spray application, and Application as a Refractory Coating (similar to Shotcrete). The thermal spray method is applied by melting or semi-melting MgO powder at high temperatures and spraying it onto the surface. The thermal spray method has the advantages of creating a protective layer resistant to high temperatures, being chemically inert, providing resistance to corrosion and being able to obtain a solid coating similar to ceramics. However, it has the disadvantages of the high melting point of MgO and therefore requiring advanced techniques such as plasma spraying and requiring controlled cooling to prevent the sprayed layer from cracking. Spray application in slurry form consists of mixing MgO powder with water and binders and applying this suspension with a spray gun. The slurry spray method has advantages such as low equipment costs and being used for refractory surface repair or internal coating. However, it has disadvantages such as requiring drying and subsequent sintering (high-temperature cooking) and affecting the quality of the application in terms of the binder selection and particle size. In the application method of Application as a Refractory Coating, MgO can be sprayed in a similar way to refractory concrete. It is especially preferred in industrial chimneys requiring a high temperature and chemical resistance (e.g., cement, metallurgy). In addition, in MgO coating applications, special attention should be paid to cleaning and roughening the surface before spraying, selecting the appropriate binder (e.g., silicates, alumina additives) for the good adhesion of MgO to the surface, ensuring slow cooling against the risk of cracking after the application, and preventing MgO from reacting with moisture and turning into $Mg(OH)_2$, which causes volumetric expansion and cracking [34–36]. Study [37] investigates the gas–liquid reaction of a $Mg(OH)_2$ suspension with SO_2. The effects of different operating parameters are investigated and a mass transfer model is developed. Th experimental results show that gas–film diffusion is the controlling mechanism of the desulfurization process. The article [38] focuses on the microstructure and biodegradable corrosion properties of Mg/MgO composite coatings by high-velocity suspension flame spraying (HVSFS). It is especially important for biomedical and environmentally friendly engineering applications.

The aim of this study is to fill a gap in the literature on the retrofitting of industrial steel chimneys with nanocoatings (MgO), which has been identified in the reviewed studies, and to compare the dynamic responses of such structures before and after reinforcement with nanocoatings. For this purpose, an industrial steel chimney with a height of 20 m, a diameter of 1 m and a thickness of 0.01 m was chosen for the reinforcement application. First, a three-dimensional (3D) finite element model (FEM) of the chimney was developed in SAP2000 software [39] using existing solid elements [40–42]. The 3D finite element model of the selected chimney has been validated using studies in the literature. Then, the FEM model of the steel chimney was analyzed using the ground motion of the 1995 Kobe earthquake, in which a large number of industrial facilities were severely damaged. Afterwards, the chimney body was reinforced with 0.001 m thick MgO to cover the entire outer body from bottom to top, and dynamic analyses were performed with the same ground motion mentioned above. Finally, chimney dynamic responses before and after MgO retrofitting, such as displacement, von Mises stress values, frequency and mode shapes, were compared.

2. Description of the Steel Chimney

An industrial steel chimney with a total height of 20 m was selected for the dynamic analysis before and after MgO retrofitting. The outer diameter of the chimney is 1.00 m. The wall thickness of the chimney is 0.01 m. The general structural properties of the chimney are shown in Figure 1.

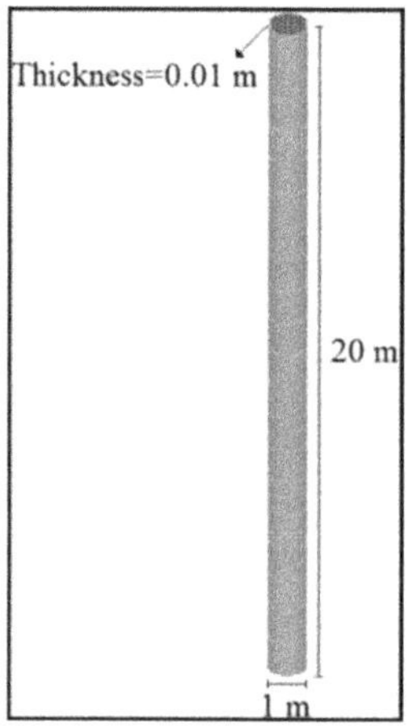

Figure 1. Dimensions of steel chimney.

3. Finite Element Modeling of the Chimney

SAP2000 software [29] was used to create three-dimensional finite element models of the chimney before and after retrofitting with MgO. This software is used for linear, non-linear, static and dynamic analysis of 3D structures [43–47]. In modeling, analytical solutions yield accurate results as long as the mesh size is properly specified. Convergence studies were carried out to determine the best mesh size in steel chimney models. The 3D FEM of the selected chimney was verified using the studies in the literature. In the finite element modeling of the chimney model, a four-node shell element is used. Each shell element has nine nodes, and each node has six degrees of freedom, namely translations in the x, y and z axes and rotations about the same axes in Figure 2.

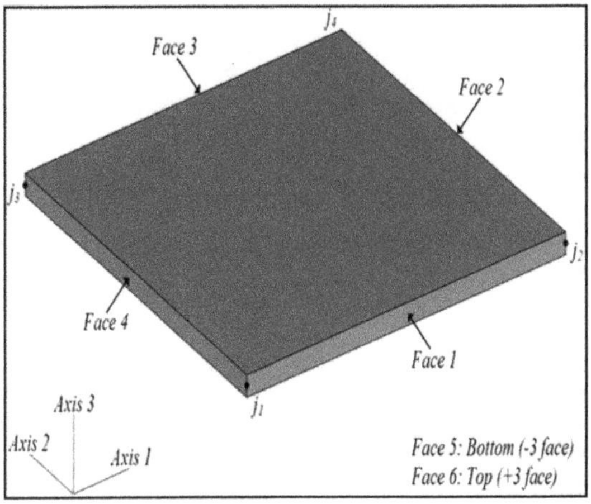

Figure 2. A schematic view and properties of the shell element [38].

Figure 2 depicts the shape and node positions for this element type. A layered shell element was used to model being coated with MgO. This element allows numerous different material layers, with different orientations and orthotropic material properties in each layer. This element can be used to model layered thick shells or solids. A schematic view of the shell element is shown in Figure 2. For thin and tall structures like chimneys, the characterization of material properties and boundary conditions, which must be taken into account in the analytical study, is critical. In the analytical model, the linear elastic material properties of steel and MgO were assumed. The chimney was built out of steel. For steel, SAP2000 software requires input data for materials such as modulus of elasticity, Poisson's ratio and unit weight. In Table 1, the elasticity modulus, Poisson's ratio and mass per unit volume of non-alloy structural steel are given [48,49].

Table 1. Material properties considered in analysis of steel chimney.

Material	Modulus of Elasticity (kN/m^2)	Poisson's Ratio	Mass per Unit Volume (kN/m^3)
Steel	2×10^8	0.30	7.849

Nanomaterials can be used to strengthen reinforced concrete, masonry structures, steel and wood in the event of flexural and shear stress due to increased loading capacity, changes in building utilization, defect correction, seismic protection and compliance with revised standards. For the retrofitting process of the chosen chimney, MgO nanocoating with the material quality shown in Table 2 was employed [50–53].

Table 2. Typical properties of MgO nanocoating.

Material	Modulus of Elasticity (kN/m^2)	Poisson's Ratio	Mass per Unit Volume (kN/m^3)
MgO	3.3×10^8	0.36	3.560

After applying the ideal meshing size to the chimney, it was discovered that the total length of linear transient analyses and total volume were extremely high, making visualization of any output impossible. As a result, the meshing size was increased to a manageable level in order to simplify the analysis and output process. A total of 1920 shell elements and 1968 nodes were utilized in the analysis at the conclusion of these studies.

When compared to the thickness of the chimney wall (0.01 m), the thickness of the MgO retrofitting (0.001 m) was thin. In the retrofitting of the model, the outer surface of the chimney was covered with a single layer of MgO with a thickness of 1 mm. The MgO layer covered the entire chimney body from bottom to top. In the literature review, it was determined that a 1 mm thick MgO layer was sufficient for reinforcement. Tensile and compressive stiffness behaviors were considered for the elements. Also, the one-way fiber direction (zero degree) was considered in this study. All degrees of freedom at the base of the chimney are assumed to be connected to the ground, and it is assumed that the chimney is fixed to the ground. In other words, in both finite element models, the boundary conditions are assumed to be fixed at the base. The adjacent nodes between the MgO layered shell elements and the steel elements are interconnected to provide the assumption of perfect bonding. In this way, it is ensured that the shell elements (steel and MgO) share the same nodal points. In addition, the MgO layered shell elements and the steel shell elements are split in the same mesh size to achieve the perfect interaction described above. In this study, modal and seismic (time-history analysis) were the used load cases for the dynamic analysis. The finite element model of the chimney and the meshing before and after MgO reinforcement are shown in Figures 3 and 4.

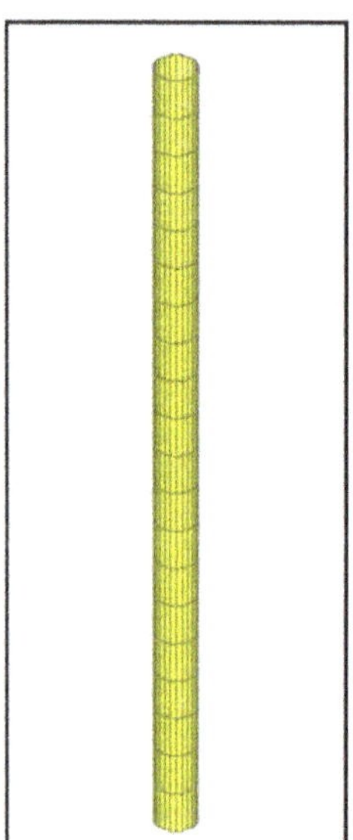

Figure 3. Finite element model of steel chimney before MgO retrofitting.

In essence, industrial chimneys are thin structures with a very homogeneous mass distribution. Thus, the higher mode effects have a significant impact on these towers' dynamic behavior. In the time-history studies, all modes having a total modal mass participation of at least 90% were taken into account. Modal analyses were carried out to determine the frequencies and mode shapes of the chimney. The modal analysis was performed with dynamic response theories. The first five frequencies of the chimney before MgO retrofitting were obtained in the range of 2.50–7.14 Hz. In Figures 5–9, the first five mode shapes and frequencies of the chimney obtained from the modal analyses are illustrated.

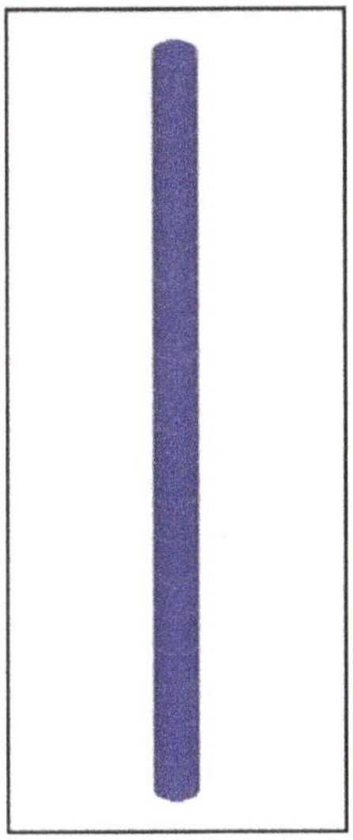

Figure 4. Finite element model of steel chimney after MgO retrofitting.

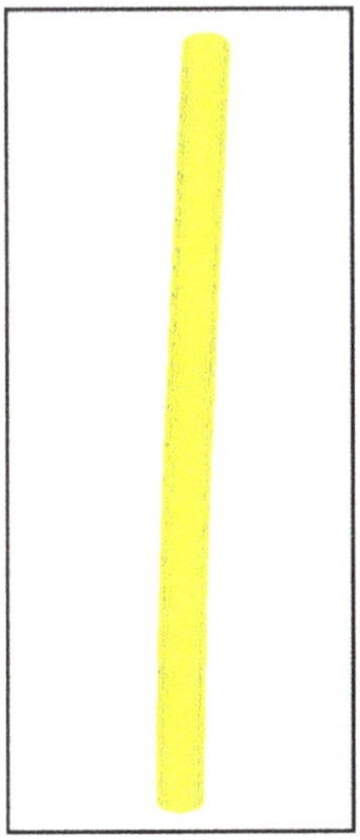

Figure 5. The 1st Mode (f = 2.50 Hz).

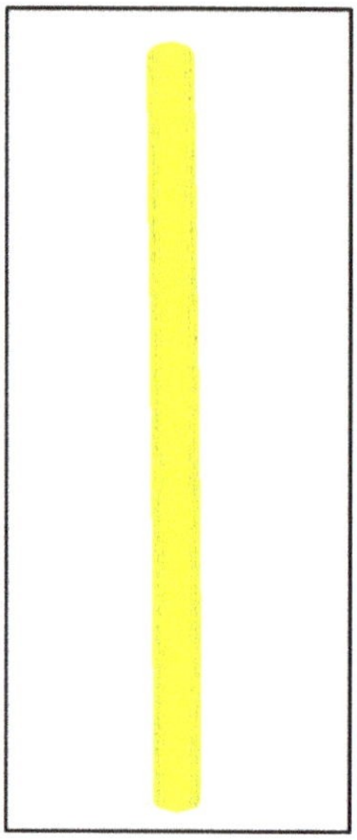

Figure 6. The 2nd Mode (f = 2.85 Hz).

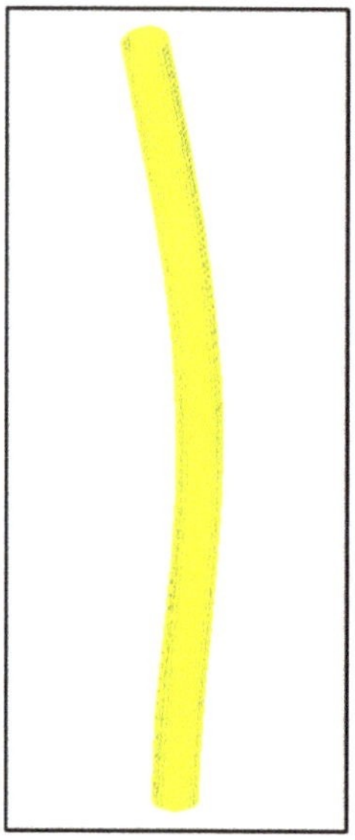

Figure 7. The 3rd Mode (f = 3.44 Hz).

The first five frequencies of the chimney after MgO retrofitting were obtained in the range of 2.63–11.11 Hz. In Figures 10–14, the first five mode shapes and frequencies of the chimney obtained from modal analyses are illustrated.

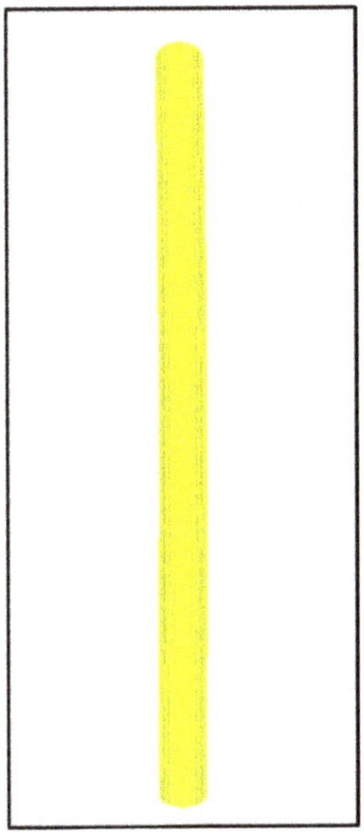

Figure 8. The 4th Mode (f = 4.54 Hz).

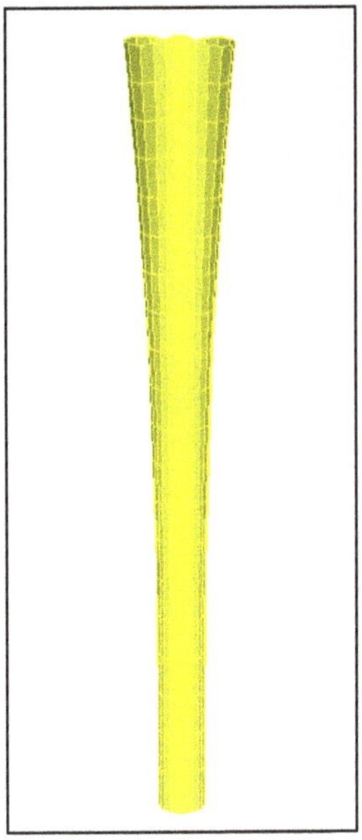

Figure 9. The 5th Mode (f = 7.14 Hz).

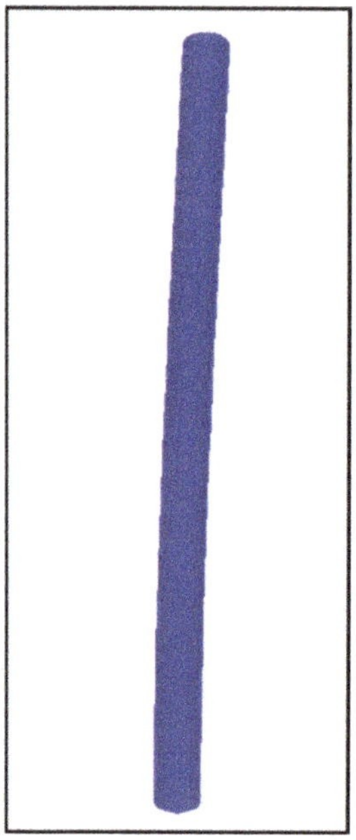

Figure 10. The 1st Mode (f = 2.63 Hz).

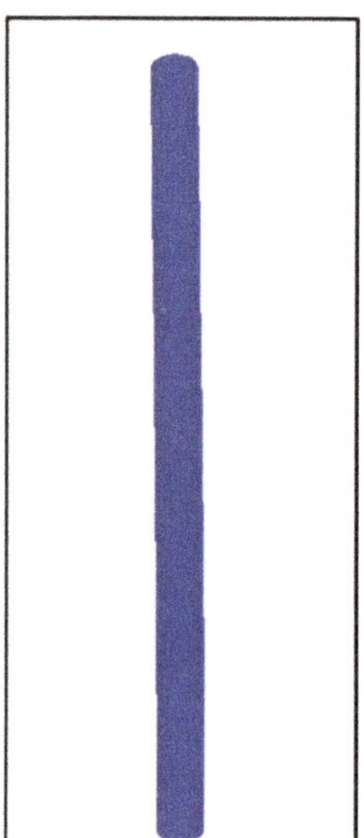

Figure 11. The 2nd Mode (f = 3.03 Hz).

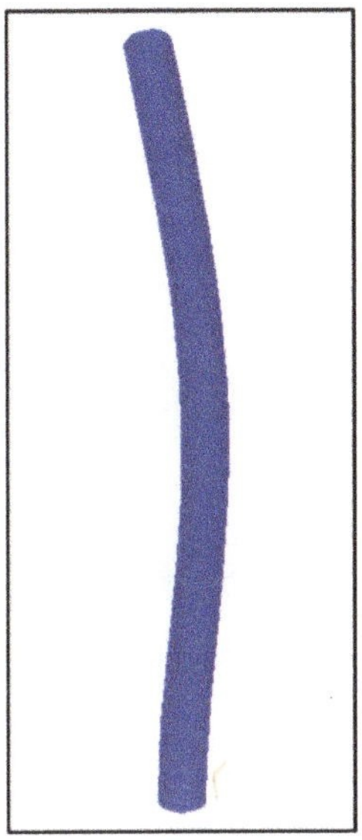

Figure 12. The 3rd Mode (f = 3.84 Hz).

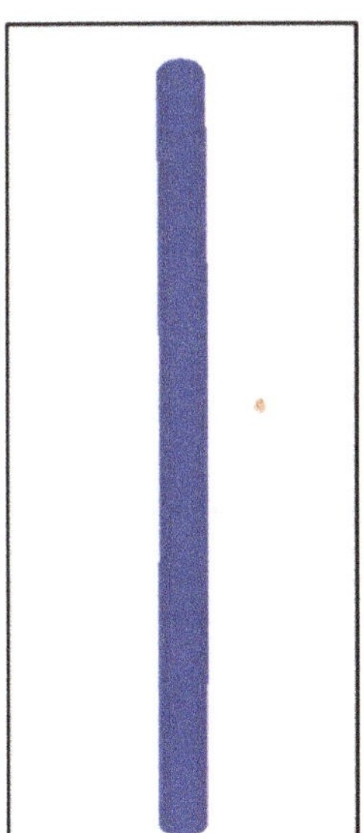

Figure 13. The 4th Mode (f = 5.55 Hz).

From Figures 5–14, it can be seen that the same mode shapes were obtained in the modal analyses before and after MgO retrofitting.

A comparison of frequency differences in the first five modes before and after MgO retrofitting is given in Table 3.

Table 3. The comparison of frequency for before and after MgO retrofitting.

Mode	1	2	3	4	5
Before MgO retrofitting (Hz)	2.50	2.85	3.44	4.54	7.14
After MgO retrofitting (Hz)	2.63	3.03	3.84	5.55	11.11
Difference (Hz)	0.13	0.18	0.40	1.01	3.97
Difference (%)	5.2	6.32	11.63	22.25	55.60

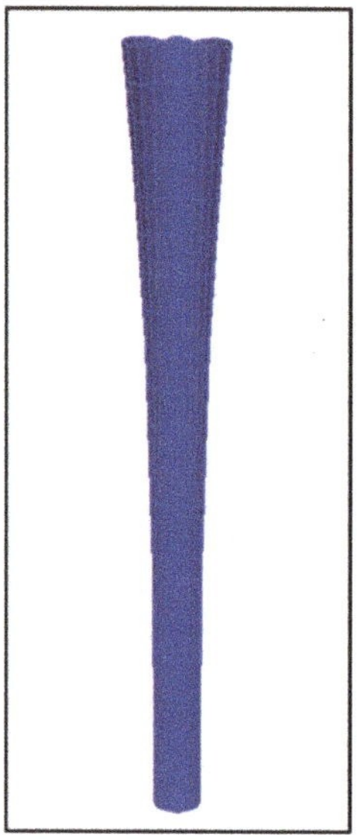

Figure 14. The 5th Mode (f = 11.11 Hz).

When the changes in natural frequencies in the first five modes were compared before and after retrofitting with MgO, it was observed that the natural frequency increased by 0.13 Hz after retrofitting in the 1st Mode. This increase corresponds to approximately 5.2% compared to the pre-retrofitting situation. When the changes in the 1st Mode shapes were examined, no change was observed in the directions of the mode shapes. In both cases, the mode shape is the translational mode. According to these findings, it is understood that there is an increase in stiffness after retrofitting.

It was observed that the natural frequency increased by 0.18 Hz after retrofitting in the 2nd Mode. This increase corresponds to approximately 6.32% compared to the pre-retrofitting situation. When the changes in the 2nd Mode shapes were examined, no change was observed in the directions of the mode shapes. In both cases, the mode shape is the translational mode. According to these findings, it is understood that there is an increase in stiffness after retrofitting.

It was observed that the natural frequency increased by 0.40 Hz after retrofitting in the 3rd Mode. This increase corresponds to approximately 11.63% compared to the pre-retrofitting situation. When the changes in the 3rd Mode shapes were examined, no change was observed in the directions of the mode shapes. According to these findings, it is understood that there is an increase in stiffness after retrofitting.

It was observed that the natural frequency increased by 1.01 Hz after retrofitting in the 4th Mode. This increase corresponds to approximately 22.25% compared to the pre-retrofitting situation. When the changes in the 4th Mode shapes were examined, no change was observed in the directions of the mode shapes. In both cases, the mode shape is the torsional mode. According to these findings, it is understood that there is a significant increase in stiffness after retrofitting.

It was observed that the natural frequency increased by 3.97 Hz after retrofitting in the 5th Mode. This increase corresponds to approximately 55.60% compared to the pre-retrofitting situation. When the changes in the 5th Mode shapes were examined, no change was observed in the directions of the mode shapes. In both cases, the mode shape is the torsional mode. According to these findings, it is understood that there is a serious increase in stiffness after retrofitting.

4. Dynamic Analyses of the Chimney

Linear transient analyses of chimney before and after MgO retrofitting were carried out using the ground acceleration values of the north–south component of the Kobe earthquake that occurred on 17 January 1995 (a total length of 50 s with equal intervals of 0.02 s). The time history of ground motion acceleration of the 1995 Kobe earthquake is given in Figure 15.

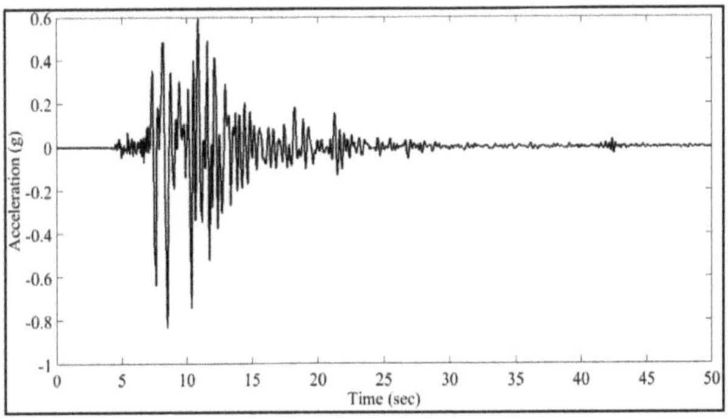

Figure 15. The time history of ground motion acceleration of 1995 Kobe earthquake (PGA = 0.833 g).

Rayleigh damping constants are calculated assuming a 2% damping ratio between the 1st Mode and the 5th Mode of the chimney [41,54–56]. Maximum horizontal displacement values under the effect of the Kobe earthquake are shown in Figures 16 and 17.

It is clearly seen in the figure that the application of MgO to the chimney reduces the maximum displacement values along the height. The maximum upper displacement value, which was 150 mm before the application of MgO to the chimney, decreased to 140 mm after the application of MgO. The reduction in maximum top displacements was about 6.66% without much additional mass to the structure. As can be seen from the contour diagrams of the maximum horizontal displacement, the maximum displacements occurred at the top of the chimney.

The maximum and minimum von Mises stress contours of the chimney exposed to the Kobe earthquake are presented in Figures 18 and 19.

As shown in Figures 18 and 19, the maximum and minimum von Mises stresses obtained before MgO retrofitting are higher than those obtained after MgO reinforcement. In other words, it is seen from Figures 18 and 19 that there is a decrease in the minimum and maximum principal stresses after MgO reinforcement. Also, there is a reduction in von Mises stresses occurring at the base of the chimney after MgO retrofitting. It can be said that MgO retrofitting is effective in reducing the minimum and maximum von Mises stresses occurring on the thin wall of the chimney.

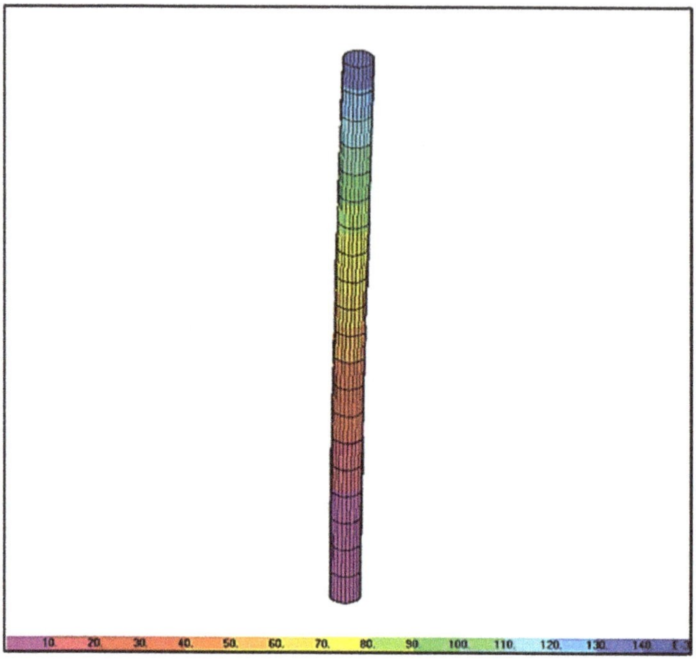

Figure 16. Maximum displacement contours of chimney before MgO retrofitting.

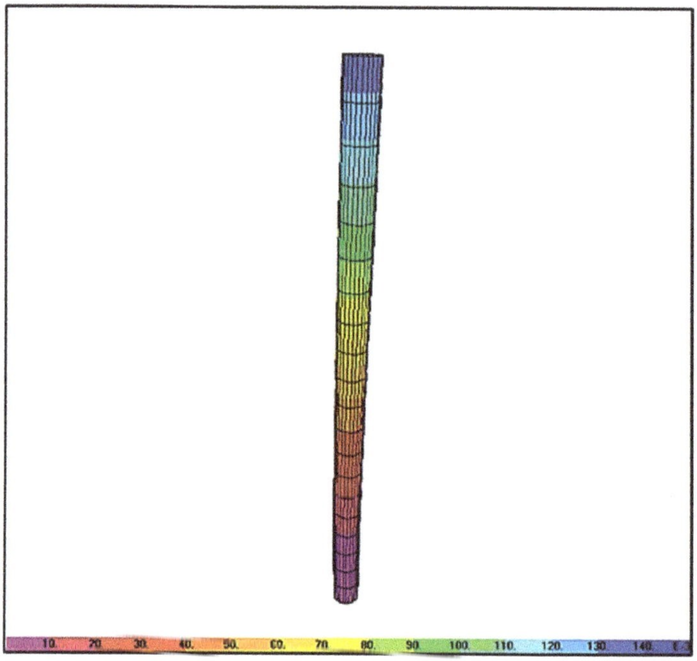

Figure 17. Maximum displacement contours of chimney after MgO retrofitting.

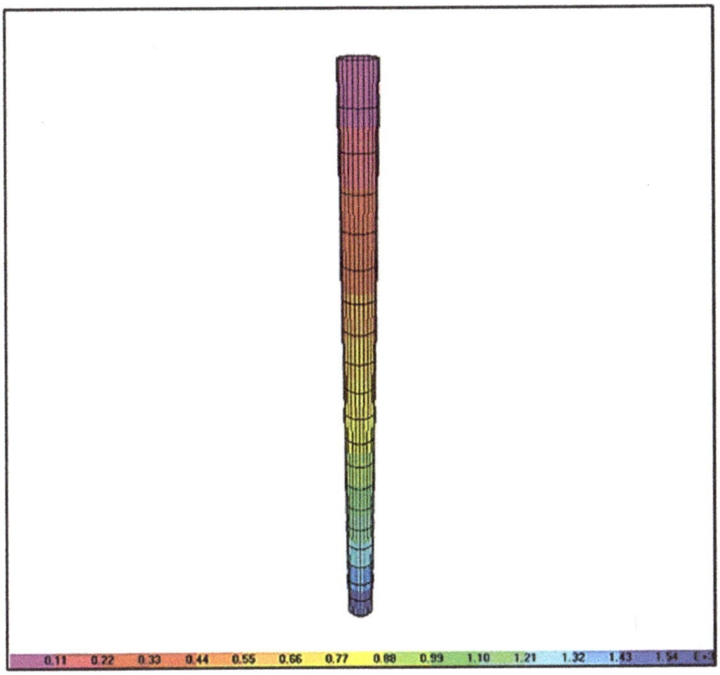

Figure 18. Minimum–maximum von Mises stress contours of chimney before MgO retrofitting.

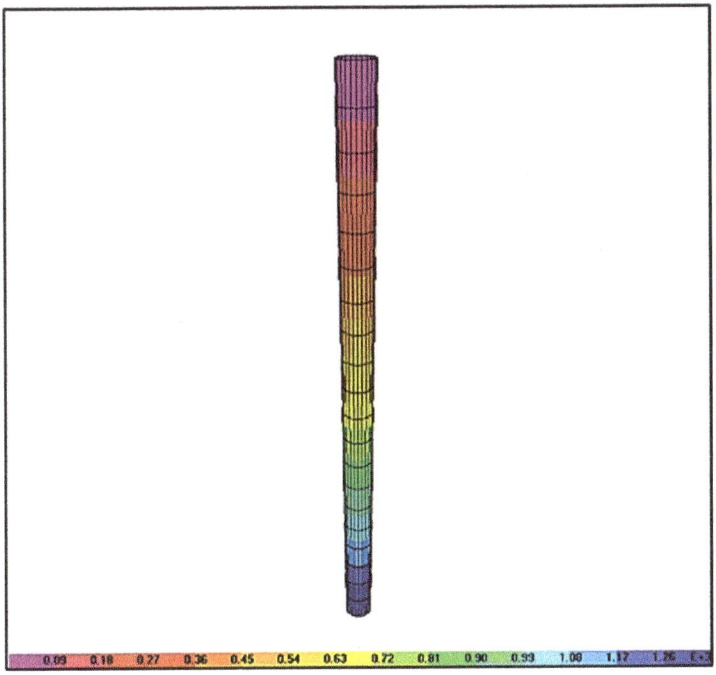

Figure 19. Minimum–maximum von Mises stress contours of chimney after MgO retrofitting.

5. Conclusions

In this study, the aim was to investigate the dynamic behavior of industrial steel chimneys before and after MgO retrofitting. A 3D finite element model of an industrial steel chimney was created in SAP2000 software and analyzed using the north–south acceleration component of the 1995 Kobe earthquake. Some of the results obtained from this study can be summarized as follows:

- From the chimney modal analysis, no changes were observed in the modal shapes before and after retrofitting with MgO. A total of five frequencies were obtained in the range of 2.50–7.14 Hz and 2.63–11.11 Hz, respectively, before and after retrofitting with MgO. It has been observed that MgO increases the rigidity of the structural system;
- For both analyses, horizontal displacement values increased from bottom to top along the height of the chimney. Maximum displacements (top) were obtained before and after MgO reinforcement at 150 and 140 mm, respectively. There was an approximate 6.66% reduction in horizontal displacement without adding too much mass to the structural system, which is a superior advantage over traditional reinforcement techniques for industrial steel chimneys;
- For both analyses, the maximum and minimum von Mises stresses formed at the bottom of the chimney. After MgO retrofitting, the minimum von Mises stress at the bottom of the chimney decreased from 37.20 kPa to 30.19 kPa, and the maximum von Mises stress from 1519.27 kPa to 1332.33 kPa. From these stress observations, it can be concluded that MgO retrofitting has an effect on stress values. Particular attention should be paid to the points where the chimneys meet the ground. With the application of MgO, the stress distribution on the chimney surface can be reduced to tolerable values without adding much additional mass to the system. This gives it a significant advantage over other traditional retrofitting techniques;
- From the displacement and von Mises stress results obtained, it can be said that MgO increases the stiffness of the structural system (increasing frequency values). More rigidity means fewer cracks; fewer cracks lead to less corrosion and stress concentrations, resulting in less maintenance and repair costs to the user;
- It can be seen from the study that nanocoatings are effective for the dynamic response of industrial steel chimneys and should be preferred as a new and good technique in the maintenance, repair and retrofitting of industrial steel chimneys. Considering the advantages of nanocoating such as a high strength/weight ratio, high corrosion resistance, easy application to all surfaces, long life and high fire resistance, it can be said that such materials will replace traditional retrofitting techniques in the near future;
- Nanocomposites are slowly becoming a popular retrofitting technique among other traditional techniques. In this respect, a large number of studies should be carried out to expand the literature on similar materials;
- In this study, a review of retrofitting a steel chimney with MgO is presented. Although the results and findings of the study belong to a specific example of steel chimney retrofitting, the methods used in this study can be applied to many steel chimneys in daily life. This study is a pioneer in demonstrating the effect of nanocoating reinforcement on the dynamic response of steel industrial chimneys;
- In future studies, experimental studies are recommended to verify the effectiveness of MgO nanocoating in field applications and to investigate long-term durability issues such as adhesion loss, environmental effects and coating degradation that may occur over time.

Author Contributions: Conceptualization, S.T.; Methodology, S.T.; Software, S.T.; Validation, F.G.; Formal analysis, F.G.; Investigation, S.T.; Resources, S.T.; Data curation, F.G.; Writing—review and editing, F.G. All authors have read and agreed to the published version of the manuscript.

Funding: This research received no external funding.

Institutional Review Board Statement: Not applicable.

Informed Consent Statement: Not applicable.

Data Availability Statement: Data are contained within the article; further inquiries can be directed to the authors.

Conflicts of Interest: The authors declare no conflicts of interest.

References

1. Karaca, Z.; Türkeli, E.; Günaydın, M.; Adanur, S. Dynamic Responses of Industrial Reinforced Concrete Chimneys Strengthened with Fiber-Reinforced Polymers. *Struct. Des. Tall Spec. Build.* **2015**, *24*, 228–241. [CrossRef]
2. Tuhta, S. OMA of Model Chimney Using Bench-Scale Earthquake Simulator. *Earthq. Struct.* **2019**, *16*, 321–327. [CrossRef]
3. Tuhta, S.; Günday, F.; Aydın, H. Subspace Identification Using N4SID Methods Applied to Model Concrete Chimney. *JournalNX* **2020**, *6*, 415–423.
4. Karaca, Z.; Türkeli, E. Determination and Comparison of Wind Loads for Industrial Reinforced Concrete Chimneys. *Struct. Des. Tall Spec. Build.* **2012**, *21*, 133–154. [CrossRef]
5. López-Patiño, G.; Adam, J.M.; Gimeno, P.V.; Milani, G. Causes of Damage to Industrial Brick Masonry Chimneys. *Eng. Fail. Anal.* **2017**, *74*, 188–201. [CrossRef]
6. Meier, U. Strengthening of Structures Using Carbon Fibre/Epoxy Composites. *Constr. Build. Mater.* **1995**, *9*, 341–351. [CrossRef]
7. Kobatake, Y. A Seismic Retrofitting Method for Existing Reinforced Concrete Structures Using CFRP. *Adv. Compos. Mater.* **1998**, *7*, 1–22. [CrossRef]
8. Zhang, S.; Ye, L.; Mai, Y.W. A Study on Polymer Composite Strengthening Systems for Concrete Columns. *Appl. Compos. Mater.* **2000**, *7*, 125–138. [CrossRef]
9. Mirmiran, A.; Zagers, K.; Yuan, W. Nonlinear Finite Element Modeling of Concrete Confined by Fiber Composites. *Finite Elem. Anal. Des.* **2000**, *35*, 79–96. [CrossRef]
10. Van Den Einde, L.; Zhao, L.; Seible, F. Use of FRP Composites in Civil Structural Applications. *Constr. Build. Mater.* **2003**, *17*, 389–403. [CrossRef]
11. Chen, C.H. Seismic Assessment of a Retrofitted Chimney by FEM Analysis and Field Testing. *Int. J. Struct. Stab. Dyn.* **2004**, *4*, 337–359. [CrossRef]
12. *ASCE/SEI 7-75*; Minimum Design Loads for Buildings and Other Structures. American Society of Civil Engineers: Reston, VA, USA, 1975.
13. Matthys, S.; Toutanji, H.; Taerwe, L. Stress–Strain Behavior of Large-Scale Circular Columns Confined with FRP Composites. *J. Struct. Eng.* **2006**, *132*, 123–133. [CrossRef]
14. Fawzia, S.; Al-Mahaidi, R.; Zhao, X.L.; Rizkalla, S. Strengthening of Circular Hollow Steel Tubular Sections Using High Modulus CFRP Sheets. *Constr. Build. Mater.* **2007**, *21*, 839–845. [CrossRef]
15. Rougier, V.C.; Luccioni, B.M. Numerical Assessment of FRP Retrofitting Systems for Reinforced Concrete Elements. *Eng. Struct.* **2007**, *29*, 1664–1675. [CrossRef]
16. Zhao, X.L.; Zhang, L. State-of-the-Art Review on FRP Strengthened Steel Structures. *Eng. Struct.* **2007**, *29*, 1808–1823. [CrossRef]
17. Pallarés, F.J.; Ivorra, S.; Pallarés, L.; Adam, J.M. Seismic Assessment of a CFRP-Strengthened Masonry Chimney. *Proc. Inst. Civ. Eng. Struct. Build.* **2009**, *162*, 291–299. [CrossRef]
18. Uomoto, T.; Mutsuyoshi, H.; Katsuki, F.; Misra, S. Use of Fiber Reinforced Polymer Composites as Reinforcing Material for Concrete. *J. Mater. Civ. Eng.* **2002**, *14*, 191–209. [CrossRef]
19. Wu, G.; Wang, H.T.; Wu, Z.S.; Liu, H.Y.; Ren, Y. Experimental Study on the Fatigue Behavior of Steel Beams Strengthened with Different Fiber-Reinforced Composite Plates. *J. Compos. Constr.* **2012**, *16*, 127–137. [CrossRef]
20. Yang, J.Q.; Smith, S.T.; Feng, P. Effect of FRP-to-Steel Bonded Joint Configuration on Interfacial Stresses: Finite Element Investigation. *Thin-Walled Struct.* **2013**, *62*, 215–228. [CrossRef]
21. Hu, L.; Feng, P.; Zhao, X.L. Fatigue Design of CFRP Strengthened Steel Members. *Thin-Walled Struct.* **2017**, *119*, 482–498. [CrossRef]

22. Kasimzade, A.A.; Tuhta, S. OMA of Model Steel Structure Retrofitted with CFRP Using Earthquake Simulator. *Earthq. Struct.* **2017**, *12*, 689–697. [CrossRef]
23. Zhou, Y.; Fan, L.; Xing, F.; Lin, W.; Hu, R. Effect of nano-SiO_2 modification on the seismic performance of recycled aggregate concrete shear walls. *Eng. Struct.* **2024**, *307*, 117945. [CrossRef]
24. Timesli, A. A Cylindrical Shell Model for Nonlocal Buckling Behavior of CNTs Embedded in an Elastic Foundation Under the Simultaneous Effects of Magnetic Field, Temperature Change, and Number of Walls. *Adv. Nano Res.* **2021**, *11*, 581–593. [CrossRef]
25. Tuhta, S. The Determination of Effect of TiO_2 on Dynamic Behavior of Scaled Concrete Structure by OMA. *Adv. Nano Res.* **2021**, *11*, 641–648. [CrossRef]
26. Tuhta, S.; Günday, F. Determination of the Effect of TiO_2 on the Dynamic Behavior of Scaled Concrete Chimney by OMA. *Mater. Technol.* **2021**, *55*, 459–466. [CrossRef]
27. Tuhta, S. The Determination of Effect of TiO_2 on Dynamic Behavior of Scaled WPC Warehouse by OMA. *Adv. Nano Res.* **2022**, *12*, 65–72. [CrossRef]
28. Günday, F. Investigation of the Effect of Coating Light Steel Container Houses with Nano-TiO_2 on Dynamic Parameters Using OMA. *Buildings* **2025**, *15*, 909. [CrossRef]
29. Krystek, M.; Górski, M. Nanomaterials in Structural Engineering. In *New Uses of Micro and Nanomaterials*; InTechOpen: London, UK, 2018. [CrossRef]
30. Satankar, R.K.; Sharma, N.; Ramteke, P.M.; Panda, S.K.; Mahapatra, S.S. Acoustic Responses of Natural Fibre Reinforced Nanocomposite Structure Using Multiphysics Approach and Experimental Validation. *Adv. Nano Res.* **2020**, *9*, 263–276. [CrossRef]
31. Erdoğan, İ.; Kısa, M.; Özen, M.; Demircan, G.; Kaya, A.İ. Cam Elyaf Takviyeli Epoksi Nanokompozitlerin Dinamik Davranışlarının İncelenmesi. *Harran Üniversitesi Mühendislik Dergisi.* **2023**, *78*, 78–90. [CrossRef]
32. Yang, H.-B.; Zhao, X.; Wang, Q.; Ruan, Y.-H.; Liu, Z.-X.; Yue, X.; Zhu, Y.B.; Wu, H.A.; Guan, Q.-F.; Yu, S.-H. Simultaneously Strengthening and Toughening All-Natural Structural Materials via 3D Nanofiber Network Interfacial Design. *Angew. Chem. Int. Ed.* **2024**, *63*, e202408458. [CrossRef]
33. Xie, T.; Han, W.; Chang, H.; Motaghedfer, M.R. Dynamic response of nanoparticle-containing concrete shells under moving loads. *Adv. Nano Res.* **2024**, *16*, 303–311. [CrossRef]
34. Ctibor, P.; Nevrlá, B.; Neufuss, K.; Petrášek, J.; Sedláček, J. Plasma Spray Coatings of Natural Ores from Structural, Mechanical, Thermal, and Dielectric Viewpoints. *Coatings* **2020**, *10*, 3. [CrossRef]
35. Shao, L.; Yu, W.; Du, X.; Shen, A.; Li, Y.; Ding, H.; Tao, S.; Wu, F. Investigating Lightweight Carbonation Curing of Waste Slurry Using Activated Magnesium Oxide: Performance Insights. *Materials* **2025**, *18*, 2084. [CrossRef]
36. Haiden, L.; Feuchter, M.; Brunner, A.J.; Barbezat, M.; Pansare, A.; Ravindran, B.; Terziyska, V.; Pinter, G. Shining a Light on Carbon-Reinforced Polymers: Mg/MgO and TiO_2 Nanomodifications for Enhanced Optical Performance. *J. Compos. Sci.* **2025**, *9*, 187. [CrossRef]
37. Shen, Z.; Ni, M.; Guo, S.; Chen, X.; Tong, M.; Lu, J. Studies on Magnesium-Based Wet Flue Gas Desulphurization Process with a Spray Scrubber. *Asian J. Chem.* **2013**, *25*, 6727–6732. [CrossRef]
38. Yao, H.-L.; Xia, J.; Yi, D.-L.; Liu, H.-Y.; Liu, W.-L.; Liu, J.-S.; Wang, F. Microstructure and Corrosion Properties of Biodegradable Mg/MgO Composite Coating on Mg Alloy Prepared by High Velocity Suspension Flame Spraying. *J. Therm. Spray Technol.* **2021**, *30*, 1544–1556. [CrossRef]
39. Computers and Structures, Inc. *SAP2000: Integrated Structural Analysis and Design Software*, Version 26.0; Computers and Structures, Inc.: Berkeley, CA, USA, 2024.
40. Aoki, T.; Sabia, D.; Rivella, D. Influence of Experimental Data and FE Model on Updating Results of a Brick Chimney. *Adv. Eng. Softw.* **2008**, *39*, 327–335. [CrossRef]
41. Kasimzade, A.A. *Finite Element Method: Foundation and Applications to Structural Analysis*; Nobel Publication: Ankara, Turkey, 2018.
42. Kasimzade, A.A. *Structural Dynamics: Theory and Applications to Earthquake Engineering*; Nobel Publication: Ankara, Turkey, 2018.
43. Sancibrian, R.; Lombillo, I.; Sarabia, E.G.; Boffill, Y.; Wong, H.; Villegas, L. Dynamic Identification and Condition Assessment of an Old Masonry Chimney by Using Modal Testing. *Procedia Eng.* **2017**, *199*, 3410–3415. [CrossRef]
44. Ke, L.; Li, C.; He, J.; Lu, Y.; Jiao, Y.; Liu, Y. Fatigue Evaluation and CFRP Strengthening of Diaphragm Cutouts in Orthotropic Steel Decks. *Steel Compos. Struct.* **2021**, *39*, 453–469. [CrossRef]
45. Kumar, Y.; Gupta, A.; Tounsi, A. Size-Dependent Vibration Response of Porous Graded Nanostructure with FEM and Nonlocal Continuum Model. *Adv. Nano Res.* **2021**, *11*, 1–17. [CrossRef]
46. Rahman, S.; Jain, A.K.; Bharti, S.D.; Datta, T.K. A Comparative Study of the Models to Predict Aeroelastic Vibrations of Circular Cylinder and Chimneys. *Wind Struct.* **2022**, *35*, 35–54. [CrossRef]

47. Günday, F. Structural Health Monitoring of Steel Garage Model with Stochastic Subspace Identification–Covariance Variance Analysis. *Buildings* **2025**, *15*, 944. [CrossRef]
48. *EN 1993-1-1;* Eurocode 3: Design of Steel Structures—Part 1-1: General Rules and Rules for Buildings. CEN: Brussels, Belgium, 2005.
49. *EN 10025-2;* Hot Rolled Products of Structural Steels—Part 2: Technical Delivery Conditions for Non-Alloy Structural Steels. CEN: Brussels, Belgium, 2019.
50. CRC. *Handbook of Chemistry and Physics*, 97th ed.; Haynes, W.M., Ed.; CRC Press: Boca Raton, FL, USA, 2016.
51. Nair, A.K.; Gautam, A.; Buehler, M.J. Nano-mechanical behavior of MgO nanostructures. *J. Appl. Phys.* **2010**, *107*, 093510. [CrossRef]
52. Walker, J.S.; Mould, S.T. Poisson's ratio of magnesium oxide: Experimental and theoretical approaches. *Phys. Rev. B* **1991**, *44*, 1434. [CrossRef]
53. Chen, Y.; Feng, Z.; Zhang, W. Effect of MgO Content on Mechanical Properties of Directionally Solidified Pure Magnesium. *Mater. Res.* **2021**, *24*, 2. [CrossRef]
54. Chopra, A.K. *Dynamics of Structures: Theory and Applications to Earthquake Engineering*, 5th ed.; Pearson: Boston, MA, USA, 2017.
55. *EN 1998-1:2004;* Eurocode 8: Design of Structures for Earthquake Resistance—Part 1: General Rules, Seismic Actions and Rules for Buildings. European Committee for Standardization: Brussels, Belgium, 2004.
56. *ASCE/SEI 7-16;* Minimum Design Loads and Associated Criteria for Buildings and Other Structures. American Society of Civil Engineers: Reston, VA, USA, 2016.

coatings

MDPI

Article

Enhanced Fracture Toughness of WC-CoCr Thermally Sprayed Coatings by the Addition of NiCrFeSiBC and Mo and Its Influence on Sliding Wear Behavior

José de Jesús Ibarra [1,*], Marco Aurelio González [2], Eduardo Rodríguez [3], Gabriel Israel Vásquez [1], Ariosto Medina [4], José Bernal [5], Claudio Aguilar [6] and Eduardo Enrique Velez [7]

[1] TecNM, Instituto Tecnológico José Mario Molina Pasquel y Henríquez, Unidad Académica Zapopan, Camino Arenero No. 1101, Colonia el Bajío, Zapopan 45019, Mexico
[2] Centro Universitario de Ciencias Exactas e Ingenierías, CONAHCYT-Universidad de Guadalajara, CUCEI, Ciudad Universitaria, Blvd. Marcelino García Barragán 1421, Guadalajara 44430, Mexico
[3] Departamento de Ingeniería de Proyectos, Centro Universitario de Ciencias Exactas e Ingenierías, CUCEI, Universidad de Guadalajara, Ciudad Universitaria, Blvd. Marcelino García Barragán 1421, Guadalajara 44430, Mexico
[4] UMSNH, Instituto de Investigación en Metalurgia y Materiales, Edificio U, Ciudad Universitaria, Morelia 58000, Mexico
[5] Instituto Tecnológico de Orizaba, Tecnológico Nacional de México, Ote. 9, Emiliano Zapata, Orizaba 94320, Mexico
[6] Departamento de Ingeniería Metalúrgica y Materiales, Universidad Técnica Federico Santa María, Valparaíso 2391206, Chile
[7] Departamento de Ingeniería Mecánica Eléctrica, Centro Universitario de Ciencias Exactas e Ingenierías, CUCEI, Universidad de Guadalajara, Blvd. Marcelino García Barragán # 1421, Guadalajara 44430, Mexico
* Correspondence: jose.ibarra@zapopan.tecmm.edu.mx

Abstract: Wear is a major issue in industry, particularly with metal components. Therefore, it is crucial to investigate methods that offer increased resistance to this phenomenon. In this research, three coating systems (pure WC-CoCr and WC-CoCr/NiCrFeSiBC+Mo, 88:12 and 83:17 wt.%) were thermally sprayed on an AISI 1018 steel substrate through the High-Velocity Oxygen Fuel (HVOF) process. The coatings were characterized using a field emission scanning electron microscope (FESEM) equipped with the energy dispersive spectroscope (EDS) and X-ray diffractometry (XRD). An analysis of the wear rate for ball-on-flat linear reciprocating sliding tribological tests for the coatings was also carried out. The coating microstructure presents well-dispersed NiCrFeSiBC splats. The WC-CoCr/NiCrFeSiBC+Mo, 88:12, system has the highest wear resistance, decreasing by 30.2% at high loads compared to commercial WC-CoCr CERMETs, and also exhibits the highest fracture toughness. Analysis of wear tracks shows that the material removal at all charges occurred mainly by an abrasive wear mechanism.

Keywords: NiCrFeSiBC; sliding wear; coatings; fracture toughness; WcCoCr; HVOF

1. Introduction

In the metal–mechanical industry, the components of any machine are exposed to different types of wear: erosive, abrasive, and sliding [1,2]. In this context, CERMET (ceramic–metallic) coatings have been demonstrated to be an excellent anti-wear option, reducing material loss considerably by this type of damage and increasing the lifespan of the components at a reduced cost of fabrication [3–5]. CERMET commonly consists of hard ceramic particles embedded in a metallic matrix; Tungsten–Carbides WC-Co and WC-CoCr are the most used to refurbish engineering components like, for example, internal gate valves, metal working tools, mining tools, and blast-spray nozzles [6]. Commonly, by using thermal spraying processes such as High-Velocity Oxygen Fuel (HVOF) and Air Plasma Spray, these materials can be applied as coatings [4,7–10]. In this sense, HVOF, it is well known, can produce more dense coatings due to the influence of its high-speed jet. However, during thermal spraying, a decarburization process occurs invariably, affecting, to some degree, the fracture toughness of the final coating due to the formation of a more brittle carbide W_2C [11–13].

On the other hand, the coatings produced with these compounds have demonstrated excellent resistance under friction wear conditions. Decarburization of W_2C decreases hardness and fracture toughness, promoting greater damage when they experience more severe operating conditions [14–17]. In this context, J. Garcia et al. [14] investigated the relationship between the microstructure and mechanical properties (hardness, modulus of elasticity, and toughness) of systems like WC-Co, SiC, Alumina, satellites, cast iron, and steel. They found that WC-Co coatings showed higher elastic modulus and compressive strength than the other materials. However, their fracture toughness was lower than that observed for cast iron and hardened steels.

Recently, some research has been conducted on enhancing fracture toughness for this type of compound. These investigations include adding reinforcements like carbon nanofibers, mixing them with metallic alloys with higher toughness, or increasing the Co binder [18,19].

Concerning the tribological performance of WC coating CERMETs, Rachidi et al. [20] evaluate the addition of 40% of a metallic alloy to produce a NiCrBSi-60%WC coating. They fabricated their samples using oxyacetylene flame thermal spraying, subjecting them to sand/wheel dry abrasive and continuous motion pin-on-disc wear tests. They concluded that NiCrBSi improves resistance to abrasive wear, and the low fracture toughness of these coatings was one of the factors that promoted the detachment of WC particles through crack propagation.

According to this, producing CERMET mixed coatings with a deformable phase that participates not as a binder alloy but as a different matrix remains challenging. Capable of improving fracture toughness to some extent without sacrificing the hardness considerably. As mentioned above, some functionally graded WC-Co or CoCr/NiCrFeSiBC higher than 20% have been studied and demonstrated an increase in fracture toughness. Still, unfortunately, under friction wear conditions, all coatings displayed lower wear resistance compared with the 100% WC, mainly due to the considerable reduction in hardness. Based on the above, the objective of this project was to produce a composite CERMET/metal coating that possesses better fracture toughness and hardness values close to that displayed by the 100% WC-CoCr coating. As a secondary material matrix, the alloy commercially known as Dimalloy 2001 (NiCrFeBSiC) was used because of its good corrosion resistance. The Molybdenum (Mo) was incorporated into this alloy to improve corrosion resistance and lubricity. Two percentages of a premixed alloy NiCrFeSiBC+Mo (12 and 17 wt.%) were added to the WC-CoCr CERMET to produce corresponding coatings through the HVOF process. Then, they underwent sliding wear tests at three different loads. Also, a detailed description of the interaction and relation of microstructure, hardness, and fracture toughness on wear factors such as wear micro-mechanism is addressed.

2. Materials and Methods

2.1. Materials and Coatings Fabrication

An AISI 1018 steel plate of 50 mm × 40 mm × 7 mm was used as a substrate for coating deposition. Subsequently, surface conditioning was performed through sand-blasting using aluminum oxide as abrasive media (particle size 250–400 μm), obtaining a final average Ra roughness of ~4.5 μm, and applying acetone cleaning to remove remaining organic contaminants and dust. Two powders were used as raw materials, WC-CoCr (Co 10%, C 5.3%, Cr 4.0, W balance) and NiCrFeSiBC+Mo (Cr 16.5%, Fe 4.5%, Si 4.2%, B 3.1%, C 0.8%, Ni balance. + Mo 2%). A 98 (NiCrFeSiBC)-2 (Mo) (wt.%) powder mix was produced through mechanical milling in a planetary ball mill (PQ-N2, Across International), using 8YSZ balls as milling media and with a 1:22 powder: ball ratio at 300 rpm after six hours of milling. Then, a homogeneous modified powder mix of WC-CoCr/NiCrFeSiBC+Mo at 88:12 and

83:12 wt.% concentrations (Figure 1a,b). Then, in the facilities of the company SURESA S.A DE C.V, by using the HVOF thermal spraying process equipped with a Jet Kote gun (See Table 1 for projection parameters), the following coating systems were fabricated: (1) WC-CoCr, (2) WC-CoCr/NiCrFeSiBC+Mo (88:12 wt.%), and (3) WC-CoCr/NiCrFeSiBC+Mo (83:17 wt.%).

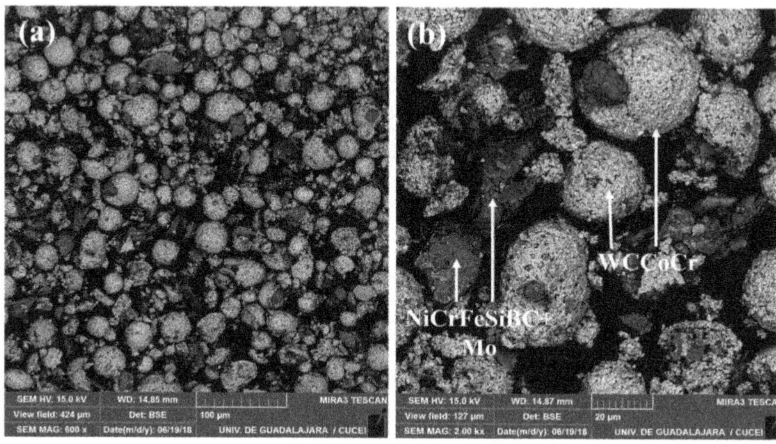

Figure 1. Micrograph of WC-CoCr/NiCrFeSiBC+Mo (83:17 wt.%) feedstock powder, (**a**) Low magnifications (600×), (**b**) High magnifications (2000×).

Table 1. HVOF parameters.

	Hydrogen Pressure (psi)	Oxygen Pressure (psi)	Argon Pressure (psi)	Oxygen Flow (%)	Hydrogen Flow (%)	Spray Distance (mm)	Feed Rate (rpm)
HVOF	110	118	105	60	80	190	3

2.2. *Coatings Characterization*

The coatings were sectioned, encapsulated in epoxy resin, roughed, and polished using diamond lap films and polish suspension. After metallographic preparation, the samples were placed in an ultrasonic bath (for 10 min for each sample), thus ensuring their cleanliness (eliminating polishing and grinding residues). Then, the identification of the splats corresponding to each powder, their distribution in the coating, and their chemical composition were determined using a Field emission scanning electron microscope (FESEM, Tescan MIRA3 LMU) equipped with a Bruker XFlash 6130 dispersive energy spectrometer, manufactured by Tescan Group, Kohoutovice Czech Republic. The identification of crystallographic phases was made using an X-ray Panalytical Empyrean diffractometer and the following testing parameters: CuKα radiation (λ = 1.542 Å) at an acceleration potential of 20 kV, a scattering (2θ) angle range from 20° to 90° and a step size of 0.02° (2θ). The microhardness of the coatings was measured on conditioned mirror-polished cross-sections using a Future-Tech FM800 hardness tester manufactured by Future-Tech Corp, Kawasaki, Japan, and by applying a load of 0.5 Kgf. Indentations were obtained using the same apparatus to determine fracture toughness (Equation (1)) by applying a 1 Kgf load and the Palmqvist method [21], which relates the cracks formed in the corners with the indentation diagonals.

$$K_{IC} = 0.0319\left(\frac{P}{a\sqrt{l_a}}\right) \quad (1)$$

where P is the load used during indentation, a is the length of the diagonal from the center to the vertex where the fracture begins, l_a is the result of subtracting the length from the center of the indentation to the end of the crack c minus the length a (see Figure 2).

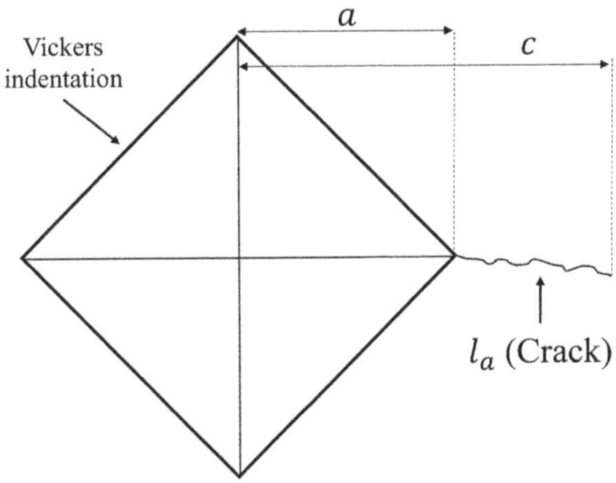

Figure 2. Palmqvist method scheme.

2.3. Sliding Wear Tests

According to ASTM G133-22 [22], tests were carried out. A contact profilometry was used to determine the coupon's average roughness (Ra) before testing, measuring a roughness between 0.02 and 0.05 μm. Subsequently, non-lubricated reciprocating wear tests were carried out at room temperature using a CTR UMT2R tribometer (Bruker, Billerica, MA, USA and applying 10, 30, and 60 N loads with a sliding stroke of 10 mm at a 3 Hz frequency with a total time of 7200 s. Aluminum balls (Al_2O_3) of 5 mm diameter were used as a counterbody. The coefficient of friction was measured in samples at 30 N and 60 N different loads.

2.4. Characterization of Worn Surfaces

The specific wear rate (W) was calculated by measuring the material lost (V) using a Profilm3D optical profilometer. The obtained value is divided by the load (F) and multiplied by the total sliding distance of the test (L) using Equation (2) [20].

$$W = \frac{V}{FL} \tag{2}$$

Finally, all worn surfaces were analyzed through scanning electron microscopy (SEM) to identify and understand the wear mechanisms involved.

3. Results and Discussion

3.1. Microstructure Characterization

Initially, specimens were properly identified, as shown in Table 2, according to their composition and load test. Figure 3a–c shows the coating microstructure of typical WC-CoCr and samples with 12% and 17% NiCrFeSiBC+Mo (dark gray splats), respectively. In addition, a certain percentage of porosity was observed ($\leq$0.43%, see Table 2). In the case of samples B-0 and C-0, the presence of NiCrFeSiBC+Mo was observed in the form of particles or continuous inter-splat films, as shown in Figure 3b,c. Subsequently, to corroborate that the dark gray splats are NiCrFeSiBC+Mo, several EDS analyses were carried out in these regions, and through an EDS spectrum, the presence of these elements could be confirmed, see Figure 3d. The thickness of the coatings was approximately 230 μm.

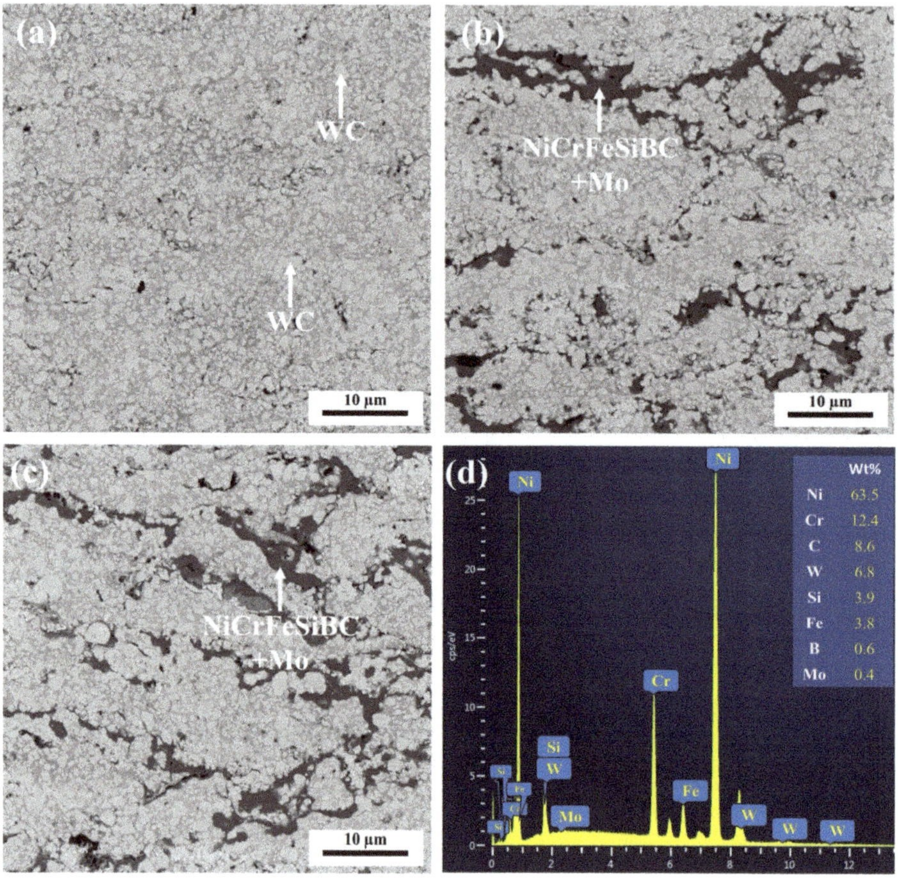

Figure 3. SEM cross-section images of samples: (**a**) A-0, (**b**) B-0, (**c**) C-0, and (**d**) EDS of NiCrFeSiBc+Mo.

Table 2. Specimens identification.

Specimens	Identification	Load (N)	Porosity	Microhardness, HV_{500}	K_{IC} ($MPa \cdot m^{1/2}$)
WC-CoCr	A-0	0	0.43 ± 0.17	1052.06 ± 40.12	12.48 ± 2.52
WC-CoCr and 12% NiCrFeSiBC+Mo	B-0		0.4 ± 0.04	887.04 ± 15.96	15.90 ± 2.83
WC-CoCr and 17% NiCrFeSiBC+Mo	C-0		0.23 ± 0.05	812.82 ± 41.25	13.61 ± 1.34
WC-CoCr	A-10	10			
WC-CoCr and 12% NiCrFeSiBC+Mo	B-10				
WC-CoCr and 17% NiCrFeSiBC+Mo	C-10				
WC-CoCr	A-30	30			
WC-CoCr and 12% NiCrFeSiBC+Mo	B-30				
WC-CoCr and 17% NiCrFeSiBC+Mo	C-30				

Table 2. *Cont.*

Specimens	Identification	Load (N)	Porosity	Microhardness, HV_{500}	K_{IC} ($MPa \cdot m^{1/2}$)
WC-CoCr	A-60				
WC-CoCr and 12% NiCrFeSiBC+Mo	B-60	60			
WC-CoCr and 17% NiCrFeSiBC+Mo	C-60				

Furthermore, Figure 4 shows the phases identified by X-ray diffraction in the sprayed coatings. In the three as-sprayed systems, WC and W_2C were identified [23–25]. During the thermal spraying, decarburization of the WC particles occurred, and the W_2C phase was formed. This process occurs according to the following reaction: $2WC \rightarrow W_2C + C$, the free carbon of this reaction gets oxidized, giving rise to CO or CO_2, which causes the decarburization of the WC particles [23]. In addition, the Co phase can be observed with less intensity [4]. On the other hand, the γ-Ni phase was identified in samples B-0 and C-0 because Ni is the element in the greatest quantity in NiCrBSiC [26,27]. The only notable difference between these two samples is the slight intensity increase in the γ-Ni peak of the C-0 sample.

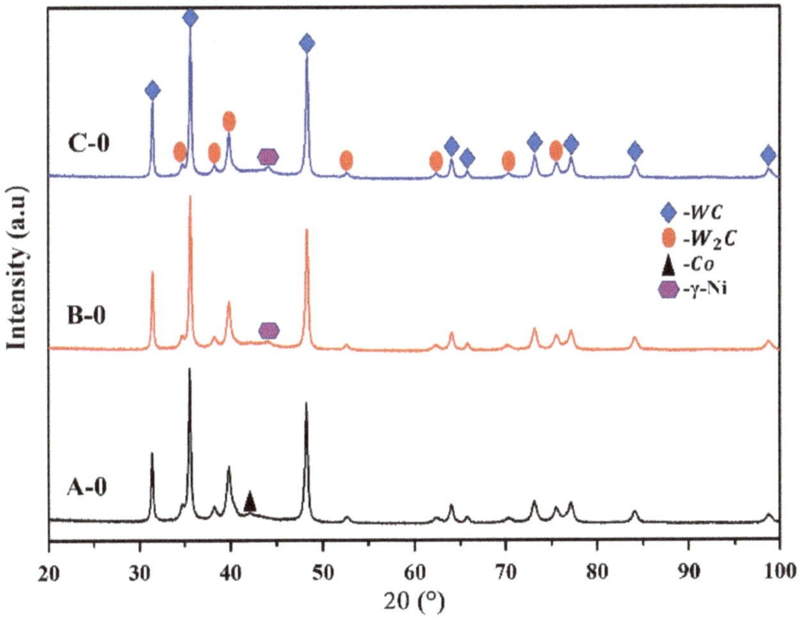

Figure 4. XRD diffractograms of as-sprayed samples.

3.2. Microhardness Fracture Toughness and Sliding Wear

The microhardness values of samples in the as-sprayed condition were measured along the cross-section of the coating microstructures (see Table 2). As expected, the maximum microhardness corresponds to sample A-0 [25,28,29]. It was found that the hardness values for the B-0 and C-0 systems decreased between 15.6% and 22.7%, with an increase in metal binder content of 12% and 17% by weight, respectively. This behavior has been observed in previous studies [27,30]. Some of the factors that influence the hardness of this type of coating are the retained WC phase, the hardness of the binder phase, and the microstructure of the coating (porosity, grain size) [31].

From the results of Table 2, the following is established: the samples with higher hardness (A-0) exhibit a reduction in their fracture toughness property compared to the B-0 and C-0 coatings. The lower number of defects and smaller area of fragile regions due to the dissolution of WC results in greater fracture toughness [31]. The fracture mechanism is governed by the unstable propagation of pre-existing defects that may be inherent to the processing or induced in the formation process [5,6]. Some cracks around each indentation exhibit different lengths and do not form in the corners of the indentations; generally, the state of stresses in these regions is higher. Also, Figure 5 illustrates the representative indentation marks resulting from K_{IC} evaluation in our systems. The upper left insets of Figure 5a–c show how the cracking path occurs, preferably along the WC/Co interface. Regarding the indentations obtained to evaluate the K_{IC}, it is evident that the crack length is shorter in sample B-0 (see Figure 5b), which presented the highest resistance to fracture toughness (K_{IC}), compared to samples A-0 and C-0 (see Figure 5a,c), which exhibit lower resistance.

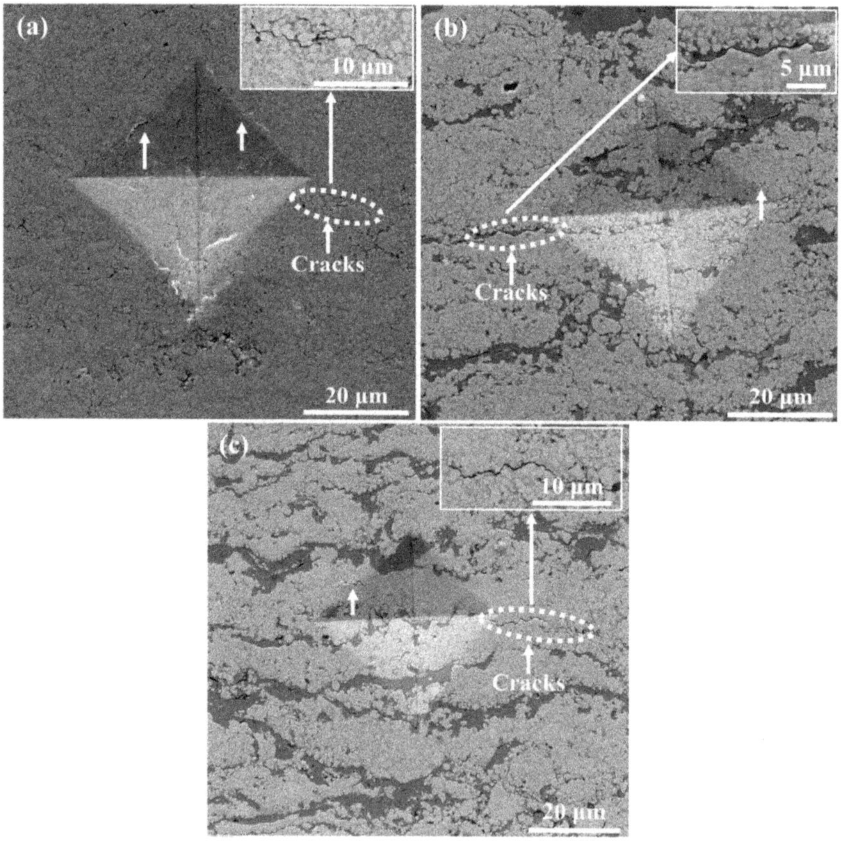

Figure 5. Micrographics of Vickers indentations of samples: (**a**) A-0, (**b**) B-0, and (**c**) C 0.

As a function of the applied load, the specific wear rate for all coating is presented in Figure 6. The average specific wear values remain relatively close for all samples at low loads (10 N), deviating somewhat when subjected to higher loads (i.e., 30 and 60 N). Usually, in brittle materials like ceramics, the rate of material loss caused by fracture is greater than that of plastic deformation [20]. The wear rate of sample B-60 decreased by 30.2% compared to the A-60. The volume wear of a brittle material depends on both the hardness and fracture toughness of the worn material and on the hardness of the abrasive [32]. Generally, the harder and tougher the matrix is, the lower the wear rate is.

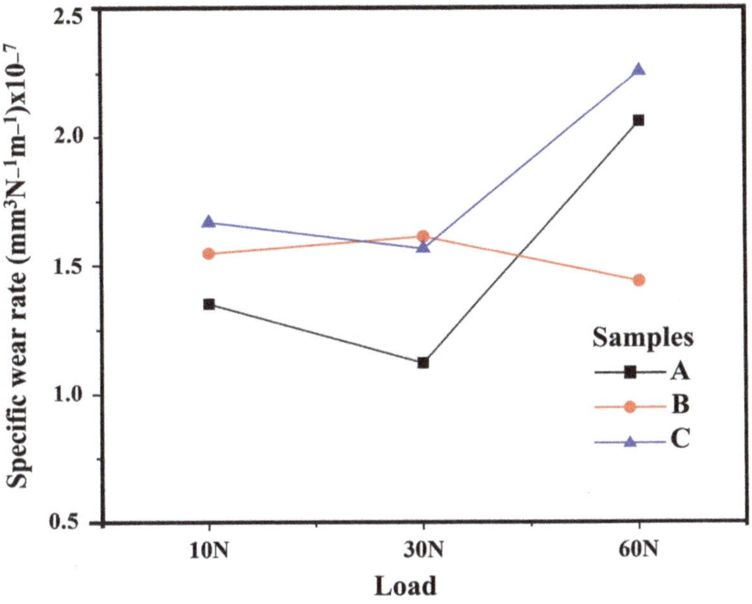

Figure 6. Specific wear rate.

3.3. *Coefficient of Friction and Wear Mechanisms*

The coefficient of friction was measured on the samples after sliding wear (loads of 30 N and 60 N), and the results are shown in Figure 7. The behavior of the coefficient of friction differs for each system according to the NiCrFeSiBC+Mo content and the load. The most significant coefficient of friction changes occurred at the highest load (60 N); see Figure 7b. Incorporating high NiCrFeSiBC+Mo contents lowers the COF significantly. While at 12%, the high elastic modulus NiCrFeSiBC+Mo particles resist deformation under the high Hertzian contact load. The lowest coefficient of friction observed at higher loads (60 N) is attributed to the formation of a more stable tribolayer directly related to the increase in temperature on the tribo-system; its stability is reflected in the spallation process delay [33]. Therefore, the coating and counter body's wear mechanism may differ.

The oxidation of elemental species within the test couples leads to the formation of tribo-films [28]. The oxidation process occurs due to the ignition temperature during the asperity interaction in the sliding contact. In WC-Co coatings on sliding contacts, the formation of tribo-films of WO_3 and CoO caused by the isothermal oxidation of WC-Co or even $WCoO_4$ at higher temperatures (~200–300 °C) has been reported [28,34]. In this regard, the formation of $CoWO_4$ and WO_3 on the worn surface of the coatings acts as a solid lubricant and reduces the friction coefficient [30,35]. Furthermore, the brittle fracture mechanism increases friction forces by providing an additional mechanism for energy dissipation in the test regions [30]. The above counteracts the lubricating effect of the oxides formed by the tribochemical processes.

SEM micrographs of the wear traces produced from the sliding wear tests are shown in Figure 8. Small pores can be visible on the worn surface mainly in samples A-30, A-60, and C-60 (see red circles in Figure 8d,g,i), likely to be sites where material pullout of WC occurred. According to the above, with a load of 60 N in the test, systems A-60 and C-60 experience greater porosity and a greater rate of wear. In this sense, stresses concentrate in the material in areas of porosity, resulting in cracks propagating from these areas [30]. This process is evidenced in samples A-30, C-30, and A-60 (see inserts of Figure 8d,g). Mechanisms of wear in samples are shown in Figure 8. The worn surfaces showed agglomeration of debris particles and parallel grooves in the sliding direction (induced by a detachment of debris particles trapped in the contact surface due to sliding). The main reason is the shear stress that resulted from normal load and friction force on the contact surface.

The presence of these visible grooves on the worn surface, formed by the passage of harder material through softer areas of the microstructure, suggests that the wear mechanism in all systems was abrasive wear. The adhesion strength between splats is further weakened due to the defects (such as porosity and micro-crack) appearing mostly on the boundaries between thin oxide sheets and splats. Later, the fatigue damage accumulates to a certain value, and micro-cracks initiate at the defects between splats. Micro-cracks grow and propagate, leading to the spallation of the splats. Thus, wear debris is generated (see inserts in Figure 8, which shows the fracture mechanism, transgranular in samples A-30, C-30, and A-60, while in B-60, the mechanism is intergranular, see yellow circles).

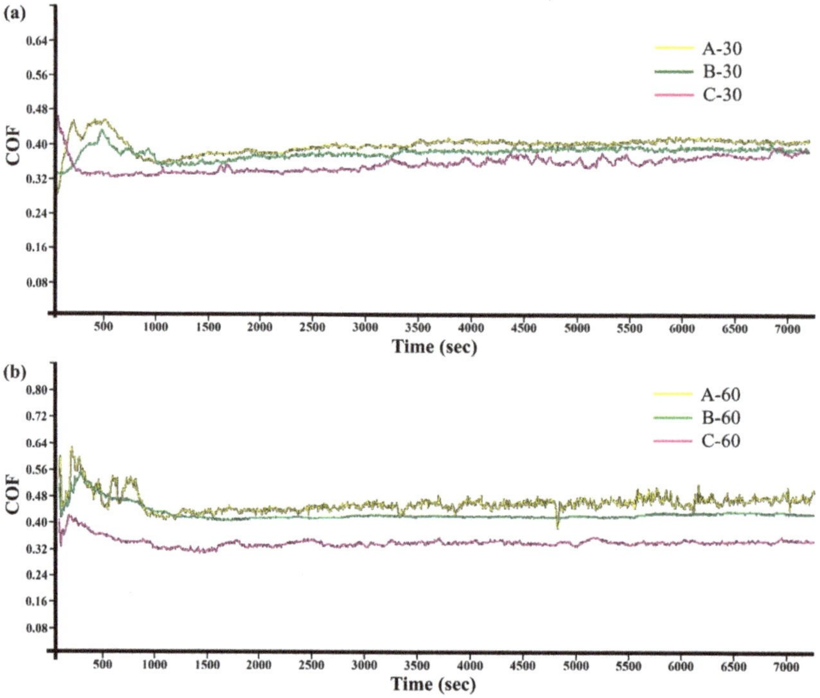

Figure 7. Coefficient of friction of samples at different loads, (**a**) 30 N and (**b**) 60 N.

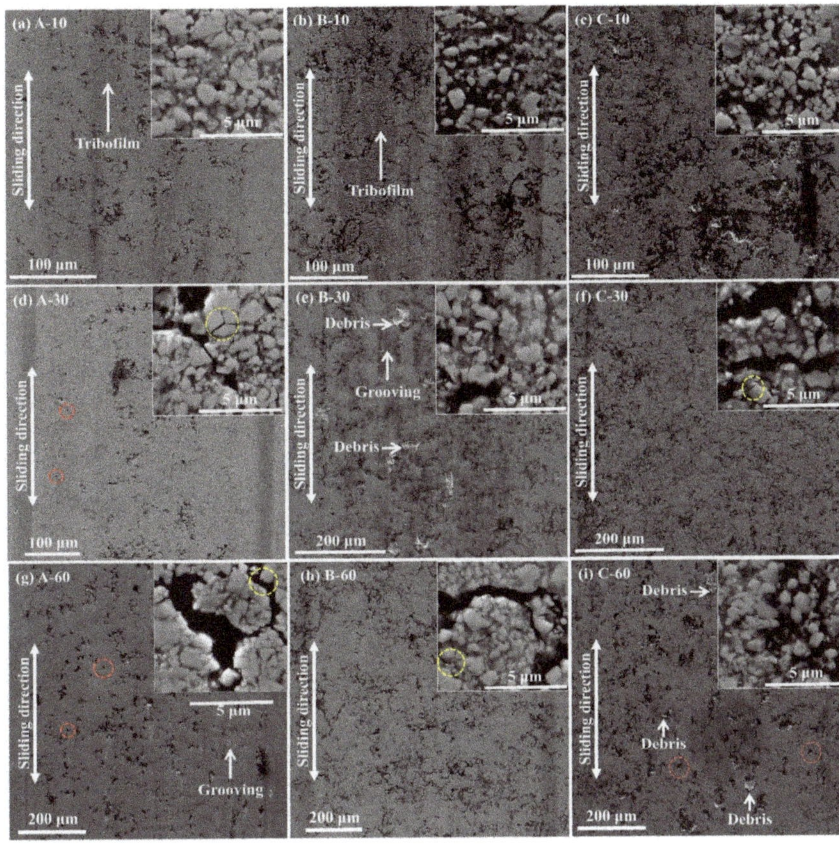

Figure 8. Mechanisms of wear, (**a–c**), (**d–f**), (**g–i**) corresponding to A, B, and C samples, respectively.

4. Conclusions

The present work demonstrated that adding 12% NiCrFeSiBC+Mo to commercial WC-CoCr CERMETS would be beneficial because the specific wear rate at higher loads (60 N) was decreased by approximately 30.2% (maintaining high hardness).

The highest resistance to fracture toughness also occurs in coatings with the adhesion of 12% NiCrFeSiBC+Mo, which is directly related to the higher wear resistance.

There is a decrease in the coefficient of friction when adhering to the WC-CoCr CERMETS (NiCrFeSiBC+Mo) in both percentages (12 and 17%) and is attributed to the formation of a more stable tribolayer directly related to the increase in temperature on the tribo-system, its stability is reflected in the spallation process delay.

Author Contributions: Conceptualization, J.d.J.I., M.A.G. and E.R.; methodology, J.d.J.I., C.A. and G.I.V.; validation, J.d.J.I., C.A. and M.A.G.; formal analysis, J.d.J.I., M.A.G. and E.R.; investigation, J.d.J.I., M.A.G. and E.E.V.; resources, M.A.G., E.R. and A.M.; data curation, G.I.V., J.B. and A.M.; writing—original draft preparation, J.d.J.I., M.A.G. and E.R.; writing—review and editing, E.R., G.I.V. and M.A.G.; visualization, G.I.V., J.B. and A.M.; supervision, J.B., E.E.V. and C.A.; project administration, J.d.J.I., M.A.G. and E.R.; funding acquisition, M.A.G., E.R. and J.d.J.I. All authors have read and agreed to the published version of the manuscript.

Funding: This research received no external funding.

Institutional Review Board Statement: Not applicable.

Informed Consent Statement: Not applicable.

Data Availability Statement: The data supporting this study's findings are available from the corresponding author upon reasonable request.

Acknowledgments: The authors thank the CONAHCYT, the Universidad de Guadalajara, and the Instituto Tecnológico José Mario Molina Pasquel y Henriquez for the technical support, and the SURESA S.A DE C.V company, Tonala, Mexico, for the deposition of the coatings.

Conflicts of Interest: The authors declare no conflicts of interest.

References

1. Ibarra, J.; Rodríguez, E.; González, M.A.; Cuenca, S.L.; Medina, A.; Vásquez, G.I. Erosion Behavior of 440C Stainless Steel Cryogenically Treated. *Rev. Mex. Ing. Quim.* **2020**, *19*, 1255–1264. [CrossRef]
2. Plucknett, K.P.; Jin, C.; Onuoha, C.C.; Stewart, T.L.; Memarrashidi, Z. The Sliding Wear Response of High-Performance Cermets. In *Handbook of Mechanics of Materials*; Springer: Singapore, 2018; pp. 1–42.
3. Zhang, S. Titanium Carbonitride-Based Cermets: Processes and Properties. *Mater. Sci. Eng.* **1993**, *163*, 141–148. [CrossRef]
4. González, M.A.; Rodríguez, E.; Mojardín, E.; Jiménez, O.; Guillen, H.; Ibarra, J. Study of the Erosive Wear Behaviour of Cryogenically and Tempered WC-CoCr Coating Deposited by HVOF. *Wear* **2017**, *376–377*, 595–607. [CrossRef]
5. Katranidis, V.; Kamnis, S.; Allcock, B.; Gu, S. Effects and Interplays of Spray Angle and Stand-off Distance on the Sliding Wear Behavior of HVOF WC-17Co Coatings. *J. Therm. Spray Technol.* **2019**, *28*, 514–534. [CrossRef]
6. Bolleddu, V.; Racherla, V.; Bandyopadhyay, P.P. Comparative Study of Air Plasma Sprayed and High Velocity Oxy-Fuel Sprayed Nanostructured WC-17wt%Co Coatings. *Int. J. Adv. Manuf. Technol.* **2016**, *84*, 1601–1613. [CrossRef]
7. Bravo, S.; Torres-González, J.; Morales-Hernández, J.; Martínez-Franco, E.; González-Olvera, J.C.; Mercader-Trejo, F.E.; Rodríguez-López, A.; Manzano-Ramírez, A.; Esparza, A.; Herrera-Basurto, R. Effect of the Manufacturing Parameters on the Quality of the Ceramic Thermal Barrier Coating after Ageing by Thermal Treatment. *Rev. Mex. Ing. Quim.* **2021**, *20*, 227–237. [CrossRef]
8. Zhou, Y.K.; Kang, J.J.; Yue, W.; Fu, Z.Q.; Zhu, L.N.; She, D.S. Wet Sliding Wear of HVOF-Sprayed WC-10Co4Cr Coatings in Simulated Seawater Drilling Fluid. *J. Therm. Spray Technol.* **2021**, *30*, 2174–2186. [CrossRef]
9. Gerner, D.; Azarmi, F.; McDonnell, M.; Okeke, U. Application of Machine Learning for Optimization of HVOF Process Parameters. *J. Therm. Spray Technol.* **2024**, *33*, 504–514. [CrossRef]
10. Federici, M.; Straffelini, G.; Gialanella, S. Pin-on-Disc Testing of Low-Metallic Friction Material Sliding Against HVOF Coated Cast Iron: Modelling of the Contact Temperature Evolution. *Tribol. Lett.* **2017**, *65*, 1–12. [CrossRef]
11. Yuan, J.; Zhan, Q.; Huang, J.; Ding, S.; Li, H. Decarburization Mechanisms of WC-Co during Thermal Spraying: Insights from Controlled Carbon Loss and Microstructure Characterization. *Mater. Chem. Phys.* **2013**, *142*, 165–171. [CrossRef]
12. Mazaheri, Y.; Khodaveysi, E.; Roknian, M.; Sheikhi, M.; Heidarpour, A. $75Cr_3C_2$-25NiCr and 86WC-10Co-4Cr High Wear- and Corrosion-Resistant Cermet Coatings Deposited on A356 Substrate by High-Velocity Oxy-Fuel Technique. *Coatings* **2022**, *12*, 1408. [CrossRef]
13. Wang, X.; Xing, Z. Preparation and Properties of Composite Nanoceramic NiCrBSi-TiO_2/WC(Co) Coatings. *Coatings* **2020**, *10*, 868. [CrossRef]
14. García, J.; Collado Ciprés, V.; Blomqvist, A.; Kaplan, B. Cemented Carbide Microstructures: A Review. *Int. J. Refract. Met. Hard. Mater.* **2019**, *80*, 40–68. [CrossRef]
15. Ahmed, R.; Vourlias, G.; Algoburi, A.; Vogiatzis, C.; Chaliampalias, D.; Skolianos, S.; Berger, L.M.; Paul, S.; Faisal, N.H.; Toma, F.L.; et al. Comparative Study of Corrosion Performance of HVOF-Sprayed Coatings Produced Using Conventional and Suspension WC-Co Feedstock. *J. Therm. Spray Technol.* **2018**, *27*, 1579–1593. [CrossRef]
16. Uzun, A.; Altuncu, E.; Ustel, F.; Turk, A.; Ozturk, S. Investigation of Wear Behaviour of HVOF Sprayed WC-Co Coatings for Automotive Parts in Different Working Conditions. *Int. J. Surf. Sci. Eng.* **2011**, *5*, 180–192. [CrossRef]
17. Stoica, V.; Ahmed, R.; Golshan, M.; Tobe, S. Sliding Wear Evaluation of Hot Isostatically Pressed Thermal Spray Cermet Coatings. *J. Therm. Spray Technol.* **2004**, *13*, 93–107. [CrossRef]
18. Pasha, M.; Kaleemulla, M. Erosive Wear Behavior of Sprayed Metal Matrix Composites: An Overview. *Proc. Inst. Mech. Eng. Part J J. Eng. Tribol.* **2013**, *227*, 1063–1075. [CrossRef]
19. Sharma, M.; Goyal, K.; Kaushal, G.; Goyal, D.K.; Scholar, R. Erosive Behaviour of HVOF Sprayed Coatings: A Review 1. *Int. J. Eng. Sci.* **2017**, *63019*, 219–243.
20. Rachidi, R.; Kihel, B.E.; Delaunois, F.; Vitry, V.; Deschuyteneer, D. Wear Performance of Thermally Sprayed NiCrBSi and NiCrBSi-WC Coatings under Two Different Wear Modes. *J. Mater. Environ. Sci.* **2017**, *8*, 4550–4559. [CrossRef]
21. Spiegler, R.; Schmauder, S.; Sigl, L.S. Fracture Toughness Evaluation of WC Co Alloys by Indentation Testing. *J. Hard Mater.* **1990**, *1*, 147–158.
22. *ASTM G133 22*; Standar Test Method for Linearly Reciprocating Ball-on-Flat Sliding Wear. ASTM International: West Conshohocken, PA, USA, 2022; pp. 1–6.
23. Shu, W.; Deng, X.; Guo, W.; Shi, W.; Li, S.; Zhang, B.; Bai, J.; Cui, Q. Microstructure and Wear Resistance of HVOF Sprayed WC-10Co-4Cr Coating on Ti-6Al-4V. *Heat Treat. Surf. Eng.* **2022**, *4*, 70–75. [CrossRef]

24. Xie, Z.; Zhang, C.; Wang, R.; Li, D.; Zhang, Y.; Li, G.; Lu, X. Microstructure and Wear Resistance of WC/Co-Based Coating on Copper by Plasma Cladding. *J. Mater. Res. Technol.* **2021**, *15*, 821–833. [CrossRef]
25. Pulsford, J.; Venturi, F.; Pala, Z.; Kamnis, S.; Hussain, T. Application of HVOF WC-Co-Cr Coatings on the Internal Surface of Small Cylinders: Effect of Internal Diameter on the Wear Resistance. *Wear* **2019**, *432–433*, 202965. [CrossRef]
26. Buytoz, S.; Ulutan, M.; Islak, S.; Kurt, B.; Nuri Çelik, O. Microstructural and Wear Characteristics of High Velocity Oxygen Fuel (HVOF) Sprayed NiCrBSi-SiC Composite Coating on SAE 1030 Steel. *Arab. J. Sci. Eng.* **2013**, *38*, 1481–1491. [CrossRef]
27. Psyllaki, P.; Vardavoulias, M.; Kekes, D.; Psyllaki, P.; Vardavoulias, M. Wear Micro-Mechanisms of Composite WC- Co/Cr-NiCrFeBSiC Coatings. Part I: Dry Sliding. *Tribol. Ind.* **2014**, *36*, 361–374.
28. Ahmed, R.; Ali, O.; Berndt, C.C.; Fardan, A. Sliding Wear of Conventional and Suspension Sprayed Nanocomposite WC-Co Coatings: An Invited Review. *J. Therm. Spray Technol.* **2021**, *30*, 800–861. [CrossRef]
29. Bergant, Z.; Batič, B.Š.; Felde, I.; Šturm, R.; Sedlaček, M. Tribological Properties of Solid Solution Strengthened Laser Cladded NiCrBSi/WC-12Co Metal Matrix Composite Coatings. *Materials* **2022**, *15*, 342. [CrossRef]
30. Pulsford, J.; Venturi, F.; Kamnis, S.; Hussain, T. Sliding Wear Behaviour of WC-Co Reinforced NiCrFeSiB HVOAF Thermal Spray Coatings against WC-Co and Al_2O_3 Counterbodies. *Surf. Coat. Technol.* **2020**, *386*, 125468. [CrossRef]
31. Nahvi, S.M.; Jafari, M. Microstructural and Mechanical Properties of Advanced HVOF-Sprayed WC-Based Cermet Coatings. *Surf. Coat. Technol.* **2016**, *286*, 95–102. [CrossRef]
32. Sun, L.; Pan, J.; Lin, C. Wear Behavior of TiC-$MoSi_2$ Composites. *Mater. Lett.* **2003**, *57*, 1239–1243. [CrossRef]
33. García-León, R.A.; Martínez-Trinidad, J.; Campos-Silva, I.; Figueroa-López, U.; Guevara-Morales, A. Wear Maps of Borided AISI 316L Steel under Ball-on-Flat Dry Sliding Conditions. *Mater. Lett.* **2021**, *282*, 128842. [CrossRef]
34. Yang, Q.; Senda, T.; Hirose, A. Sliding Wear Behavior of WC-12% Co Coatings at Elevated Temperatures. *Surf. Coat. Technol.* **2006**, *200*, 4208–4212. [CrossRef]
35. Guilemany, J.M.; Miguel, J.M.; Vizcaino, S.; Climent, F. Role of Three-Body Abrasion Wear in the Sliding Wear Behaviour of WCCo Coatings Obtained by Thermal Spraying. *Surf. Coat. Technol.* **2001**, *140*, 141–146. [CrossRef]

coatings

MDPI

Article

Methods of Distributing the IF-WS_2 Modifier for Its Introduction into the Structure of the Al_2O_3 Aluminum Oxide Coating

Joanna Korzekwa [1], Mateusz Niedźwiedź [1], Grzegorz Dercz [1], Krzysztof Cwynar [2], Maciej Sowa [3], Marek Bara [1,*] and Wojciech Simka [3]

1 Institute of Materials Engineering, Faculty of Science and Technology, University of Silesia in Katowice, 75 Pułku Piechoty 1a, 41-500 Chorzów, Poland; joanna.korzekwa@us.edu.pl (J.K.); mateusz.niedzwiedz@us.edu.pl (M.N.); grzegorz.dercz@us.edu.pl (G.D.)

2 Institute of Chemistry, Faculty of Science and Technology, University of Silesia in Katowice, Szkolna 9, 40-006 Katowice, Poland; krzysztof.cwynar@us.edu.pl

3 Faculty of Chemistry, Silesian University of Technology, 44-100 Gliwice, Poland; maciej.sowa@polsl.pl (M.S.); wojciech.simka@polsl.pl (W.S.)

* Correspondence: marek.bara@us.edu.pl

Abstract: The microstructures and structures of modified Al_2O_3/IF-WS_2 coatings prepared on aluminum substrates are studied. Amorphous Al_2O_3 oxide coatings are obtained on EN AW 5251 aluminum alloy using the electrooxidation process. The quality of the IF-WS_2 nanopowder is of great importance in the process of its introduction into the nanopores of the Al_2O_3 oxide coating. Commercial nanopowder tends to agglomerate, and without appropriate pretreatment, it is difficult to introduce it into the nanopores of the coating. To improve the degree of fragmentation of the IF-WS_2 nanopowder, an experiment was carried out to distribute the nanopowder in the presence of strong ultrasounds, and new conditions for introducing the powder into the nanopores were used. A two-level design of experiment (DOE) was used. The SEM examination made it possible to conclude that Method A contributed to a more even distribution of nanoparticles in the microstructure of Al_2O_3 coatings. GIXD analyses showed the presence of WO_3 derived from the IF-WS_2 modifier next to crystal structures derived from aluminum and WS_2. Modification of coatings using Method A resulted in surfaces with lower contact angles measured with polar liquids and higher surface free energy compared to Method B.

Keywords: aluminum anodizing; Al_2O_3 coatings; IF-WS_2 modifier distribution; SEM and GIXD research; contact angle measurements

1. Introduction

High-strength aluminum alloys are widely used in various branches of the machinery, automotive, and aerospace industries, mainly due to their exceptional weight-dependent mechanical properties. The most frequently used alloy additives in the production of aluminum alloys are Cu, Si, Mg, Mn, and Zn. These elements influence the strength and corrosion resistance properties of aluminum alloys in different ways. For example, alloys of the 3xxx–6xxx series (i.e., the addition of Mn, Si, Mg, Mg + Si, respectively) are resistant to corrosion, while alloys of the 7xxx series (i.e., the addition of Zn + Mg) are characterized by the highest strength of all aluminum alloys, with low corrosion resistance [1]. Alloys that are also worthy of special attention are alloys containing 2 ÷ 5% Mg and 0.1 ÷ 0.4% Mn (e.g., EN AW-5251), which are characterized by good corrosion resistance, high plasticity, and weldability. An additional advantage of these alloys is the low content of intermetallic phases, which increases the susceptibility to anodic oxidation. Without proper preparation, the surface of aluminum or its alloys is also exposed to adhesive wear due to friction.

One well-known method that provides adequate corrosion protection and significantly improves surface properties is anodizing. This process, which depends on many parameters (e.g., voltage or temperature, electrolyte composition, the nature of the anodized substrate, and additional technological treatments before and after anodizing), is complex and multidisciplinary [2,3]. Research on the anodizing process of aluminum and its alloys has been ongoing for over 100 years and continues to be a current research trend, largely due to the potential for producing aluminum and its alloys from recycled sources, which is expected to significantly contribute to the reduction of CO_2 emissions [4]. Matijosius et al. demonstrated in their studies that the phase of anodic Al_2O_3 depends on the method of its formation, composition, and the nature of the substrate [5]. As demonstrated by Elkilany et al., intermetallic phases in aluminum alloys influence the growth rate and morphology of anodic coatings. In the oxide coatings formed on Al 2014 and 6061 alloys, which contained coarse intermetallic phases, significant voids and surface defects were observed. In contrast, the anodic coating formed on the 7075 alloy exhibited greater uniformity and fewer defects due to the presence of fine precipitates [6]. Based on his research, Kchaou concluded that the anodizing time has a noticeable impact on the thickness of the oxide coating. Longer anodizing times resulted in a thicker and harder aluminum oxide coating, which, in his studies, led to an increased coefficient of friction [7]. Benea et al. found that extending the duration of the oxidation process to a constant potential and consequently increasing the thickness of the nanoporous aluminum oxide coatings results in more homogeneous coatings [8]. Dervishi et al. found that the porosity and thickness of the anodic coating depend on the anodizing time and the type of electrolyte. Studies on the porosity of anodic coatings produced in phosphoric acid showed that the porosity of the coatings decreases with increasing anodizing time [9]. Kwolek et al. observed that the nanoscale porosity of the coating on 5005 aluminum alloy changes with the thickness of the coating as a result of increased anodizing voltage, significantly affecting the protective properties of the coatings [10]. As demonstrated by Brudzisz et al., the incorporation of anions during the anodization of metals plays a crucial role in the formation of anodic oxides, affecting the resulting morphology. Additionally, the incorporated anions lead to the emergence of unique features, such as photoluminescence and galvanoluminescence of anodic aluminum oxide. Therefore, anion incorporation is recognized as a method of electrochemical doping of nanostructured anodic oxides, which may advance new applications [11].

In recent decades, considerable attention has been given to various modifications aimed at improving the tribological properties of anodic coatings produced on aluminum alloys. The surface oxide coatings formed in the hard-anodizing process are materials with highly developed surfaces, and their properties can be very diverse, depending mainly on the conditions in which they are produced [12]. It has been found that the porosity of anodic oxide coatings can be utilized as a lubricant reservoir to create self-lubricating systems, reducing friction and wear [13,14]. The primary challenges were often the dimensions of the lubricant powder particles, which exceeded the pore sizes of the oxide coatings. Additionally, it was observed that a reduction in the thickness of the aluminum oxide coating resulted in a decrease in the mechanical strength of the oxide surface. Interest in the aforementioned issue was also raised in studies [15], where authors emphasized that the selection of the dopant substance is not straightforward. In recent years, nanoparticles as lubricant additives have increasingly garnered attention [16–21]. The current state of knowledge clearly indicates that nanolubricants can enhance tribological performance and processing efficiency due to their nanoscale characteristics; they minimize friction, wear, and energy consumption, thereby extending the lifespan of components [22]. Tungsten disulfide (WS_2), similar to molybdenum disulfide (MoS_2) and graphite, belongs to the category of solid lubricants. Transition metal dichalcogenides MS_2 (M = Mo, W) and their complexes are part of a large class of so-called two-dimensional or layered materials. Within the layers, atoms are bonded by strong covalent forces, whereas the individual layers are held together by Van der Waals forces [23]. Due to their layered structure and the weak Van der Waals interactions between layers, WS_2, MoS_2, and graphite are widely used to provide consistent lubrication through a mechanism of exfoliation. MS_2 (M = Mo, W) in the form of hollow onion-like nanoparticles (known as IF—inorganic fullerene-like nanostructures) were the first inorganic fullerene-like materials, discovered in 1992 [19]. Subsequent research produced IF with a more uniform, spherical shape to enhance tribological properties [24]. These lubricating substances, often referred to as lubricants, exhibit a low coefficient of friction and excellent lubricating properties, making them highly valued in technical dry sliding friction assemblies. Due to its ability to operate across a wide temperature range and under high pressure (in vacuum up to 1316 °C), WS_2 is one of the most effective solid lubricants for reducing friction in sliding components [25–27]. For instance, recent studies by Li et al. demonstrate the beneficial impact of graphene addition on the structure, corrosion resistance, and friction properties of coatings prepared by the MAO method on AA1060 aluminum. To prepare the graphene-doped coatings, graphene was dispersed in isopropanol with the aid of ultrasound and then added to the electrolyte in which the coatings were formed [28]. Studies on microtextured anodic coatings doped with nano-MoS_2, obtained using ultrasonic impregnation, have been presented by Chen et al. [29]. The significance of methods for preparing homogeneous suspensions containing nanoparticles was demonstrated by Han et al. [30].

Given the wide range of potential applications for tungsten disulfide as a lubricating material, the authors of this work, as well as previous studies, consider it prudent to explore methods for modifying metal surfaces by applying different WS_2 deposition conditions. In our recent study, we discussed the issue of nanoparticle agglomeration arising from the quality of nanopowders [31]. Therefore, it appears necessary to explore methods for breaking down agglomerates of modifiers and obtaining a homogeneous suspension to facilitate the incorporation of modifiers into nanoporous oxide coatings. This article presents the results of an experiment aimed at improving the dispersion of inorganic fullerene-like tungsten disulfide (IF-WS_2) nanopowder in the presence of strong ultrasonication using novel conditions for introducing the powder into nanopores of Al_2O_3 coatings. In the article, an attempt was also made to correlate scanning electron microscope (SEM) and grazing angle incidence X-ray diffraction (GIXD) analysis with surface wetting angle measurements and surface tension studies, which has not been found in articles of similar subject matter. The paper employs statistical techniques in the design and analysis of Design of Experiments (DOE) to optimize processes and elucidate the relationships between input variables and measured outcomes. The SEM analysis revealed that Method A facilitated a more uniform dispersion of nanoparticles within the microstructure of the Al_2O_3 coatings. GIXD analysis identified the presence of WO_3, originating from the IF-WS_2 modifier, alongside crystalline structures from aluminum and WS_2. The modification of coatings via Method A resulted in surfaces exhibiting lower contact angles with polar liquids and higher surface free energy compared to those modified using Method B.

2. Materials and Methods

2.1. Sample Preparation

The material under investigation was an Al_2O_3/IF-WS_2 composite oxide film produced via a two-step process. The first step involved the fabrication of a graded Al_2O_3 coating on the surface of the European aluminum wrought with magnesium (EN AW-5251) aluminum alloy. Prior to oxidation, the samples underwent sequential etching in a 5% KOH (Avantor Performance Materials Poland S.A., Gliwice, Poland) solution for 45 min, followed by a 10% HNO_3 (Avantor Performance Materials Poland S.A., Gliwice, Poland) solution for 10 min, both at room temperature. The oxide film was obtained through the electrochemical oxidation of the aluminum alloy in a ternary solution comprising sulfuric (Avantor Performance Materials Poland S.A., Gliwice, Poland), oxalic (Avantor Performance Materials Poland S.A., Gliwice, Poland), and phthalic acids (Avantor Performance Materials Poland S.A., Gliwice, Poland). The hard-anodizing process was performed at a constant electric charge density of 180 $A{\cdot}min/dm^2$, using varying current densities and times (Table 1) and a constant temperature of 303 $\pm$ 1 K. In the second step of the process, inorganic fullerene-like

(IF) tungsten disulfide (WS_2) nanoparticles (NPs; NanoMaterials Ltd., Yavne, Israel) were incorporated into the porous alumina coating using two distinct methods:

- The Al_2O_3-coated aluminum alloy samples were immersed in a container filled with ethanol containing IF-WS_2 NPs at a concentration of 15 g/L. This container was then subjected to sonication using a VCX 130 sonicator (Sonics & Materials Inc., Newtown, CT, USA) operating at a frequency of 20 kHz, delivering 10 kJ of energy to the system. The sonication process lasted for 5 min. Following this, the samples were left undisturbed in the solution for approximately 24 h to allow for natural settling.
- Similar to Method A, the container was initially sonicated under the same conditions. However, in the second step, the container was additionally placed in an ultrasonic bath for 15 min. After this secondary sonication, the samples were removed from the solution.

Table 1. Conditions for coating formation and WS_2 NPs incorporation, based on a two-level factorial design.

Sample Designation	The Electrolysis Conditions for Charge Density of 180 [A·min/dm^2]		The Conditions of Introducing IF-WS_2 NPs
	Time [min]	J [A/dm^2]	
1	60	3 (−1)	Method A (−1)
2	60	3 (−1)	Method B (1)
3	30	6 (1)	Method A (−1)
4	30	6 (1)	Method B (1)

The key difference between the methods is the additional ultrasonic bath treatment in Method B, which involves a second stage of sonication for 15 min after the initial treatment, while Method A relies solely on the initial sonication and subsequent settling period.

Table 1 presents the conditions for coating formation and nanopowder incorporation based on a two-level factorial design without replication.

2.2. Microscopic Analysis

To analyze the composition of the nanosized Al_2O_3 fibers containing self-arranged IF-WS_2 NPs, the fabricated samples were examined using a HITACHI S-4700 (Hitachi, Tokyo, Japan) scanning electron microscope (SEM) equipped with a NORAN Vantage digital energy dispersive X-ray microanalysis (EDS) system. The surface and cross-section of the Al_2O_3/IF-WS_2 NPs coatings were analyzed using SEM in the secondary electron (BSE) mode with a yttrium aluminum garnet (YAG) detector. The YAG single crystal was employed to minimize electron dose. The cross-section of the Al_2O_3/IF-WS_2 NPs was acquired by fresh fracture. Prior to this analysis, the samples were coated with a thin carbon film using a sputtering apparatus to enhance image contrast and prevent blurring caused by charge accumulation.

2.3. Structural Analysis

Grazing angle incidence X-ray diffraction (GIXD) experiments were conducted using an X'Pert Philips PW 3040/60 diffractometer (PANalytical, Almelo, The Netherlands), operating at 30 mA and 40 kV, equipped with a vertical goniometer and Eulerian cradle. The radiation wavelength (λCuKa) employed was 1.54178 Å. GIXD patterns were recorded within the 2θ range from 20° to 70° with a step size of 0.058 for incident angles of 0.2°, 0.3° 0.5°, and 1.0°. Lower incident angles allow X-rays to penetrate only the surface coatings of the sample, while higher incident angles enable deeper penetration into the sample [32]. To ensure comparable intensities of diffraction lines, the conditions for pattern collection (step size and counting time) were appropriately adjusted.

2.4. Surface Wettability Methodology

The surface wettability of the samples was assessed by measuring the contact angle using the sitting drop method. A video-based goniometer, OCA 15EC (DataPhysics Instruments GmbH, Filderstadt, Germany), was employed for this purpose [33,34]. Measurements were conducted at a constant ambient temperature of about 298 K. Two types of liquids were utilized: a polar liquid (distilled water) and a non-polar liquid (diiodomethane). Ten drops of each liquid were deposited along the length of each sample using a micropipette equipped with an automatic feeder. Following the application of each drop, a photograph was taken and transferred to a computer. The software employed enabled automatic measurement of the contact angle on both the right and left sides of the droplet, which were subsequently averaged. The average values of the angles obtained from the ten drops were utilized to determine the average contact angle for each sample. Two methods were employed to determine surface free energy: the OWRK method (Owens, Wendt, Rabel, and Kaelble) [35], also known as the two-factor method, which is widely used for assessing the surface free energy of solids; and the Wu method [36], also referred to as the harmonic method, which offers an alternative to the OWRK method. In the OWRK method, the surface free energy γ_s of the material is divided into two components: dispersive γ_s^d and polar γ_s^p. This approach utilizes equations that incorporate the surface energies of the liquid and solid, as well as the contact angles of the liquid on the solid surface.

The OWRK equation is as follows:

$$\gamma_s(1 + cos\theta) = 2(\sqrt{\gamma_s^d \gamma_l^d} + \sqrt{\gamma_s^p \gamma_l^p}), \quad (1)$$

where γ_s^d is the dispersive component of the surface free energy of a solid, γ_s^p is the polar component of the surface free energy of a solid, γ_l^d is the dispersive component of liquid surface free energy, and γ_l^p is the polar component of the liquid surface free energy.

To determine γ_s^d and γ_s^p, the contact angle θ must be measured for at least two liquids with known values of γ_l^d and γ_l^p.

The Wu method, similarly to the OWRK method, divides the surface free energy into two components: dispersive γ_s^d and polar γ_s^p. The difference is in how these components are calculated.

The Wu equation (i.e., the harmonic method) is as follows:

$$\gamma_s(1 + cos\theta) = 4(\frac{\gamma_s^d \gamma_l^d}{\gamma_s^d + \gamma_l^d} + \frac{\gamma_s^p \gamma_l^p}{\gamma_s^p + \gamma_l^p}), \quad (2)$$

where γ_s^d is the dispersive component of the surface free energy of a solid, γ_s^p is the polar component of the surface free energy of a solid, γ_l^d is the dispersive component of liquid surface free energy, and γ_l^p is the polar component of the liquid surface free energy.

To apply the Wu method, similar to the OWRK method, it is essential to employ at least two measuring liquids with known dispersive and polar components of surface energy. The Wu method utilizes a harmonic distribution of energy components, offering an alternative approach to analyzing the surface properties of materials.

2.5. Experimental Design

To investigate the impact of input parameters on the contact angle, an experiment based on the 2k factorial design without repetition was conducted using Statistica 13 software. The independent variables chosen for this study were current density (3 and 6 A/dm^2) and method (A and B), while the dependent variable was the contact angle. A benchmark analysis of all dependent variables against a factorial design and a two-factor model with interaction was conducted. A Pareto chart was selected due to its efficiency in identifying the factors exerting the greatest influence on the process. This chart is commonly employed to assess the magnitude and significance of effects. In statistical software, the p-value is frequently utilized for interpreting the statistical significance of factors in the process. In the Pareto chart, bars that intersect the reference line (in the current article, p-value = 0.05) are considered statistically significant at the 0.05 level. In statistical analysis, normal probability plots of residuals are utilized to assess whether the residuals of a model (i.e., the differences between observed and predicted values) follow a normal distribution. This is crucial because many statistical techniques, such as linear regression, assume that residuals are normally distributed. If the residuals are normally distributed, the points on the plot should align along a straight line. Deviations from this line may indicate departures from normality [37]. Additionally, marginal mean plots with 95% confidence intervals were utilized to visually represent the statistical analysis of the results. The DOE methodology is discussed in detail in studies [38,39].

3. Results and Discussion

Figure 1a–d shows the SEM/YAG-BSE images (magnification ×1000) of the cross-sections of Al_2O_3/IF-WS_2 coatings from samples 1 to 4. The images highlight the Al_2O_3 coating, the EN AW 5251 aluminum alloy substrate, and the embedded IF-WS_2. In Figure 1c,d, the visible boundary between the coating and the substrate is not marked, as the coating appears delaminated due to the fracturing of the sample. Figure 2a–d shows the SEM/YAG-BSE images (magnification ×1000) of the surface of an Al_2O_3/IF-WS_2 coatings from samples 1 to 4. In our view, the free settling of NPs in Method A (Figure 2a,c) results in better dispersion of the NPs across the surface, with fewer agglomerations of IF-WS_2 NPs forming within the spherical indentations of the coating compared to Method B (Figure 2b,d). The free settling method also allows for more uniform incorporation of the IF-WS_2 NPs into the upper coating of the coating (Figure 1a,c). In contrast, Method B (Figure 1b,d), which involves the introduction of NPs using an ultrasonic bath, likely facilitates the penetration of the NPs deeper into the coating. This effect is particularly evident for coatings obtained at a current density of 3 A/dm^2 (Figure 1b).

The GIXD spectra of the Al_2O_3/IF-WS_2 composite coatings for samples 1–4 are presented in Figure 3a–d. The analysis of GIXD diffraction patterns obtained in this study indicates that the polycrystalline structure with a cubic phase consists of aluminum coatings, with the (111) and (200) planes corresponding to peaks at 38.08° 2θ and 44.03° 2θ, respectively, in agreement with JCPDS PDF 5 004-0787. The detection of the crystalline aluminum phase is most likely due to the glancing interaction of the beam with the surface of the coating, leading to its contact with the edges of the cross-sectioned sample, where substrate material may be present. Alternatively, this effect could be caused by chips from the substrate introduced during the cutting of samples for GIXD analysis. Furthermore, the presence of the rhombohedral phase of WS_2 modifier (JCPDS PDF 5 04-003-4480) and the tetragonal phase of WO_3 (JCPDS PDF 5 005-0388) were detected. The observed X-ray "amorphous halo" suggests that a portion of the material exists in an amorphous state, corroborated by a broad peak centered at 28° 2θ, indicative of the amorphous Al_2O_3 oxide coating. A detailed examination of the diffraction patterns revealed variations in the contributions of crystalline and amorphous phases as the beam angle (α) increases. At the smallest angle of α = 0.20°, corresponding to the surface coating, the crystalline phases (Al and WS_2) dominate. With an increase in the beam angle to α = 1.00°, the amorphous phase becomes more pronounced. Based on diffraction analysis obtained by the GIXD method, it was found that the obtained material has a layered structure, where the upper zone is a mixture of amorphous, two crystalline phases, and the WO_3 phase in two cases. In the deeper layers, for the beam angle of α = 1.00°, the zone is the amorphous phase and WS_2. The reduced frequency of diffraction lines originating from WS_2 is likely due to a diminished concentration of WS_2 in the lower regions, approaching the detection limit of the method. It is important to emphasize that identifying the amorphous phase in the X-ray analysis encompasses materials lacking long-range order and nanocrystalline materials where the crystallites do not exceed 3–4 nm in size.

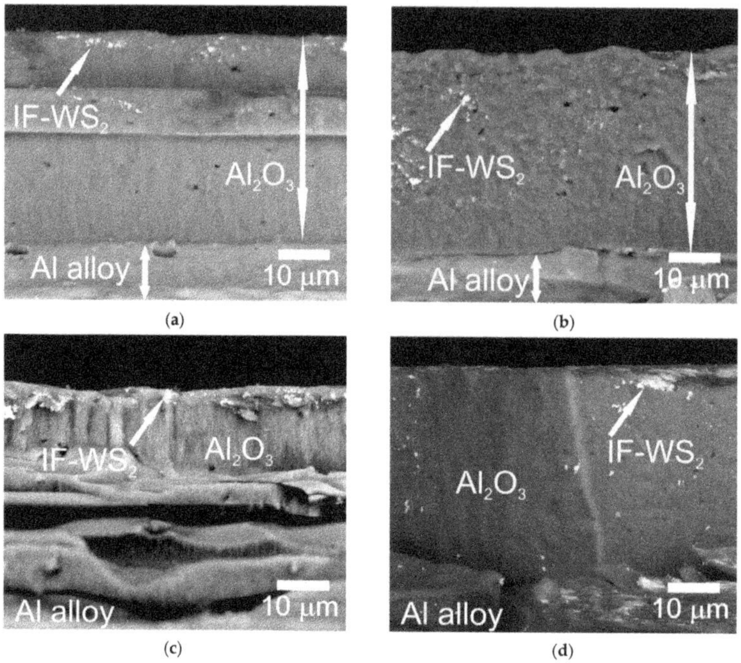

Figure 1. Microstructures (×1000) of the fresh cross-section of the Al_2O_3/IF-WS_2 coatings: (**a**) sample 1—Method A; (**b**) sample 2—Method B; (**c**) sample 3—Method A; and (**d**) sample 4—Method B.

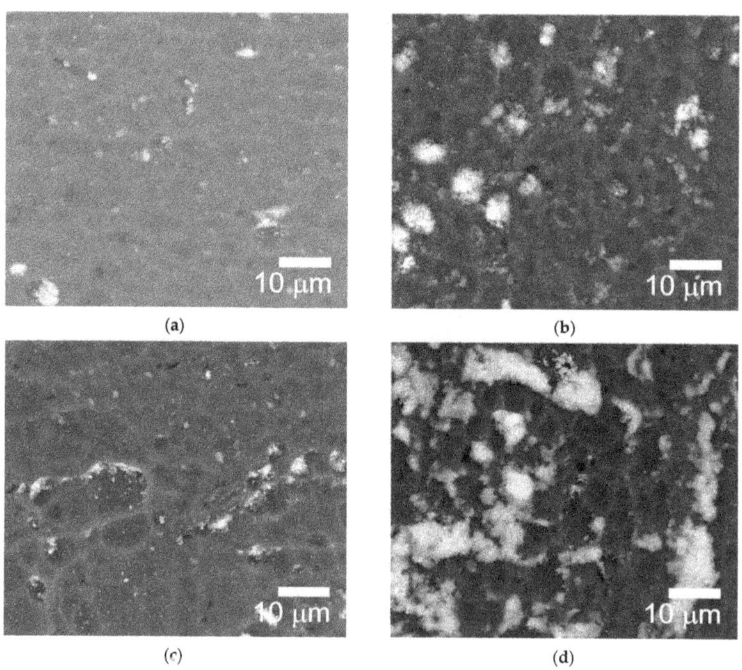

Figure 2. Microstructures (×1000) of the surface of the Al_2O_3/IF-WS_2 coatings. (**a**) sample 1—Method A; (**b**) sample 2—Method B; (**c**) sample 3—Method A; and (**d**) sample 4—Method B.

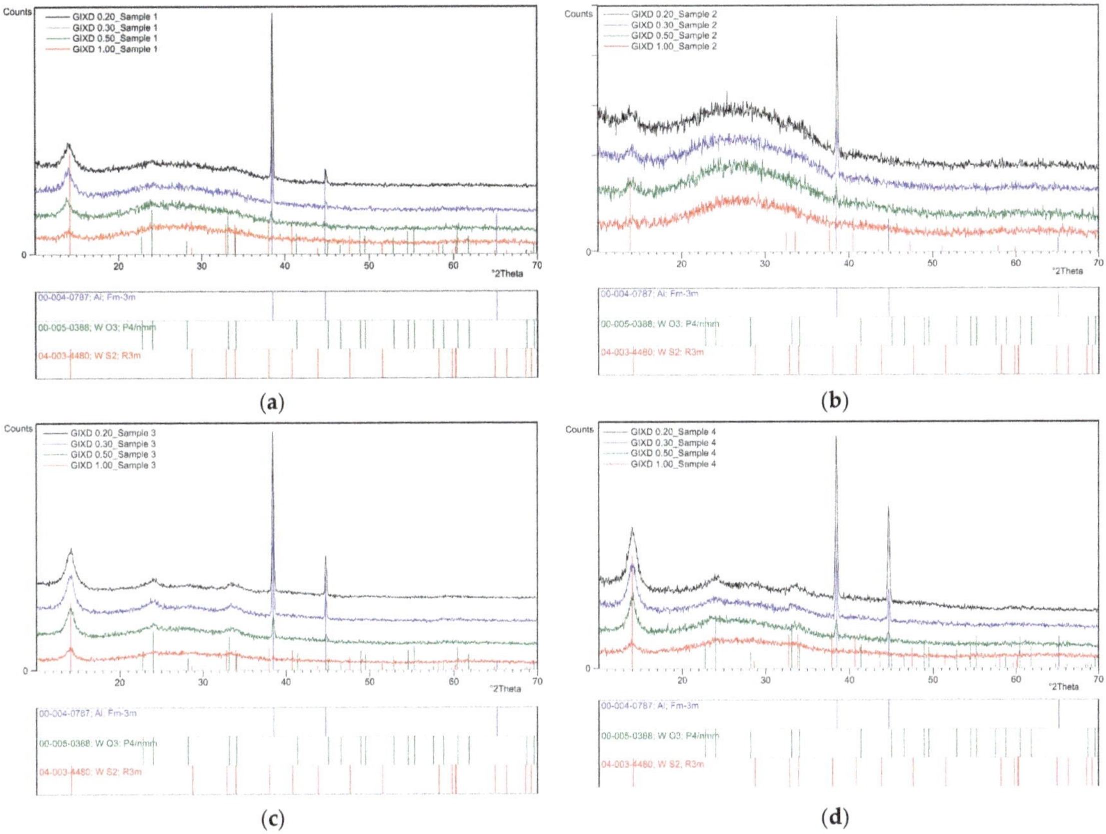

Figure 3. GIXD data of Al_2O_3/IF-WS_2 coatings: (**a**) sample 1—Method A; (**b**) sample 2—Method B; (**c**) sample 3—Method A; and (**d**) sample 4—Method B.

Figure 4 shows the summary of GIXD diffractograms obtained at an incidence angle of 0.50° for all samples. It can be observed that the highest reflections originating from the WO_3 phase are present in samples 3 and 4. The formation of WO_3 from WS_2 is most likely associated with the high energy during the dispersion of WS_2 in a sonicator. Other methods for synthesizing WO_3 were cited in Pragati's work [40], including Mandal's study, in which water-dispersed WO_3 was prepared from commercial WS_2 powder [41].

Figure 5a–d shows the SEM/YAG-BSE micrographs (magnification ×10,000) of the cross-sections of Al_2O_3/IF-WS_2 coatings from samples 1 to 4. In samples obtained by both methods, the Al_2O_3 nanofibers with IF-WS_2 nanoparticles dispersed between them are visible. A detailed examination of the fresh cross-sections of the coatings shown in Figure 5a,c reveals that the IF-WS_2 nanoparticles, visible as bright spots, are uniformly distributed within the near-surface coating of the porous alumina coating produced using Method A. In contrast, the distribution of the IF nanoparticles differs significantly in the samples prepared by Method B, as depicted in Figure 5b,c. In these samples, large agglomerates are evident on the coating's surface, with some penetrating deeply into the voids of the porous film, which is more clearly illustrated in Figure 1b,d.

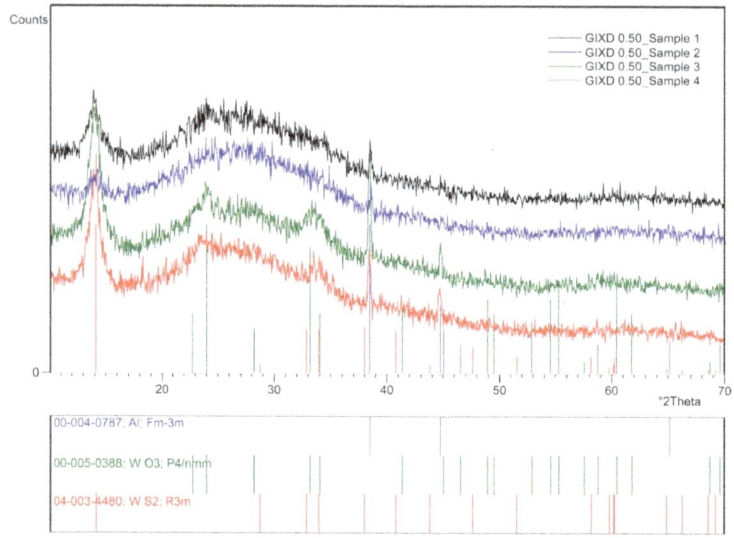

Figure 4. GIXD data of Al_2O_3/IF-WS_2 coatings for all samples at the beam angle of $\alpha = 0.50°$.

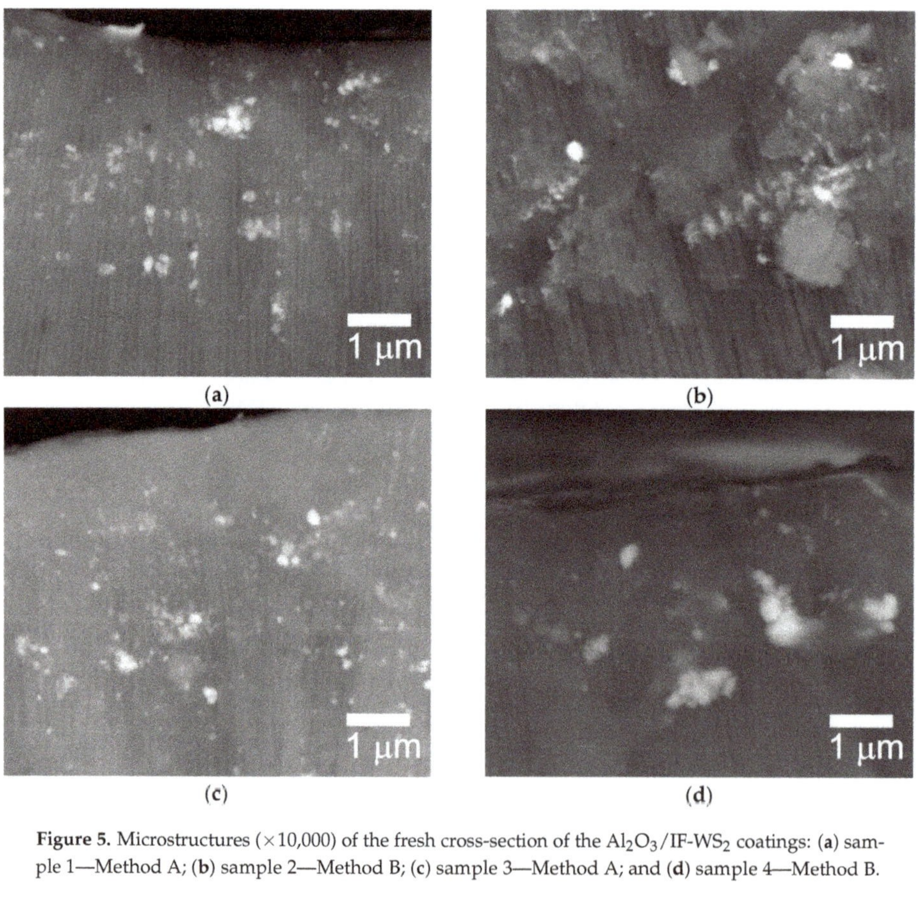

Figure 5. Microstructures (×10,000) of the fresh cross-section of the Al_2O_3/IF-WS_2 coatings: (**a**) sample 1—Method A; (**b**) sample 2—Method B; (**c**) sample 3—Method A; and (**d**) sample 4—Method B.

EDS analysis was performed at the points labeled 1 and 2 in Figure 6. The results are summarized in Table 2. The "Atom %" column indicates the atomic percentage (at%) of each element, while the "Compnd %" column represents the weight percentage (wt%). The chemical composition of the coating was determined by identifying and quantifying the elements present in the selected regions, with the results expressed in terms of their respective compound formulas.

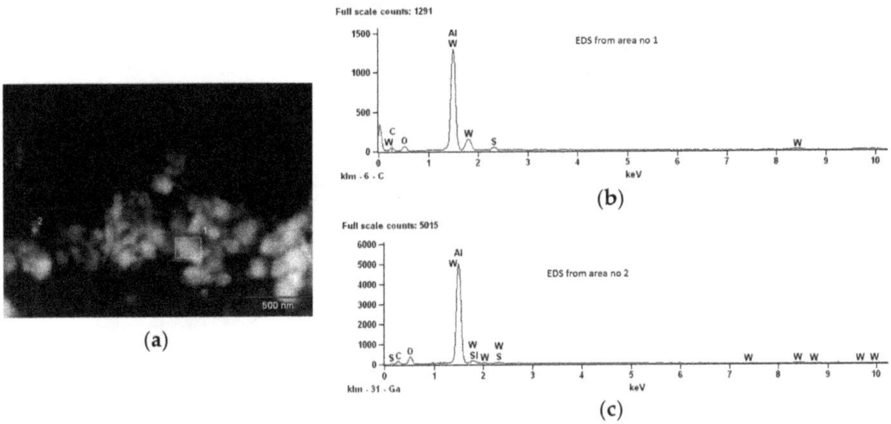

Figure 6. (**a**) SEM/YAG-BSE of a fresh fracture cross-section with a marked area of EDS analysis; (**b**) EDS from area no 1; and (**c**) EDS from area no 2.

Table 2. Quantitative results for EDS from areas number 1 and 2.

Area	Element Line	Atom %	Atom % Error	Formula	Compnd %
1	*O K*	45.44	±2.26	O	27.30
	Al K	49.30	±0.50	Al	49.94
	S K	2.38	±0.16	S	2.87
	W L	2.88	±0.25	W	19.90
	Total	100.00			100.00
2	*O K*	48.80	±1.07	O	34.05
	Al K	49.25	±0.28	Al	57.96
	S K	1.16	±0.10	S	1.63
	W L	0.79	±0.09	W	6.36
	Total	100.00			100.00

Figure 7a–d shows the SEM/YAG-BSE micrographs (magnification ×25,000) of the surface of Al_2O_3/IF-WS_2 coatings from samples 1 to 4. When analyzing the coating surfaces at higher magnifications, the difference in the deposition of IF-WS_2 nanoparticles is not as clearly visible as at lower magnifications. While observing these surfaces, one can notice fine, brightly shining spots of randomly distributed IF-WS_2 nanoparticles or aggregates formed by them.

Table 3 shows the results of contact angle measurements using distilled water and diiodomethane and surface free energy.

Table 3. Contact angles of oxide coatings measured with distilled water and diiodomethane.

Sample	Contact Angle (°) (Diiodomethane)	Standard Deviation (°)	Contact Angle (°) (Distilled Water)	Standard Deviation (°)	Surface Free Energy WORK [mN/m]	Surface Free Energy WU [mN/m]
1	51.60	2.58	65.65	2.74	39.34	45.18
2	47.64	1.79	68.15	5.14	39.27	45.07
3	57.59	4.53	65.21	3.59	38.20	43.86
4	44.39	4.28	75.39	5.76	37.95	42.83

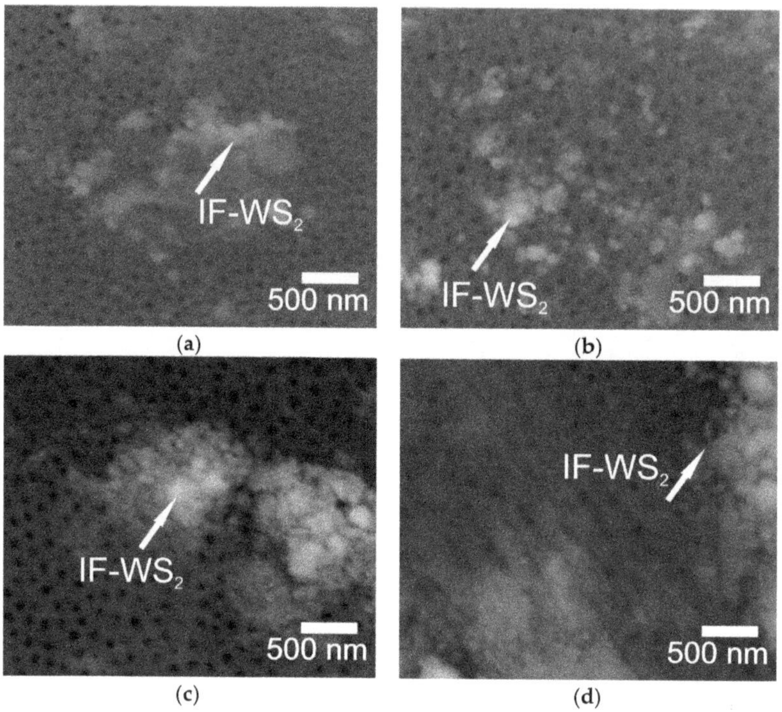

Figure 7. Microstructures (×25,000) of the surface of the Al_2O_3/IF-WS_2 coatings: (**a**) sample 1—Method A; (**b**) sample 2—Method B; (**c**) sample 3—Method A; and (**d**) sample 4—Method B.

Coatings produced at both 3 A/dm^2 and 6 A/dm^2 using Method A for applying IF-WS_2 NPs exhibit more uniform and lower contact angle measurements (Figure 8a) when using distilled water as the polar liquid (Figure 9a,e) compared to Method B (Figure 9c,g), indicating an increase in surface hydrophilicity for surfaces obtained by Method A. The surface free energy (Figure 10) is higher for samples anodized for 60 min at a current density of 3 A/dm^2, indicating that a longer anodization time combined with a lower current density promotes the formation of surfaces with higher surface energy. This could be advantageous for enhancing adhesive properties and protective coatings. Modification of the coatings using Method A results in surfaces with lower contact angles measured with a polar liquid and higher surface free energy compared to Method B, suggesting that Method A is more effective in increasing the hydrophilicity and surface energy of the coatings. This effect may be due to the uniform distribution of IF-WS_2 NPs within the near-surface oxide coating. In the case of Method B, the formation of agglomerates on the oxide coating surface significantly altered the surface wettability, particularly noticeable for sample 4. Such a method may be applicable for oxide coatings intended to minimize surface wettability.

Contact angle measurements using a non-polar liquid (diiodomethane) showed significantly lower values, primarily due to the liquid's properties, namely its much lower surface tension compared to water and substantially lower polarity. In summary, coatings produced at both 3 A/dm^2 and 6 A/dm^2 using Method A for applying IF-WS_2 NPs exhibit higher contact angles (Figure 8b) when diiodomethane is used as the test liquid (Figure 9b,f) compared to Method B (Figure 9d,g). Therefore, coatings obtained by Method A demonstrate lower surface wettability.

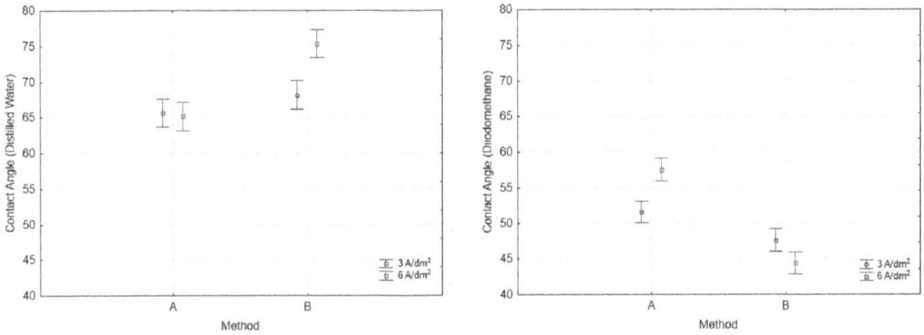

(a) (b)

Figure 8. Charts of marginal means and confidence intervals (95%) for the contact angle measured using (**a**) distilled water and (**b**) diiodomethane.

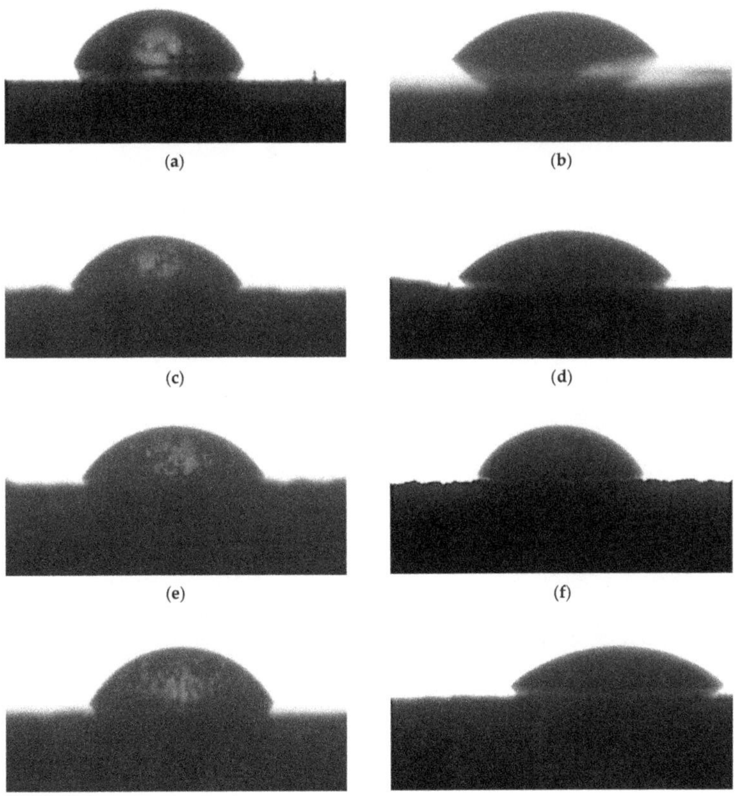

(a) (b) (c) (d) (e) (f) (g) (h)

Figure 9. Images of droplets of the media used during contact angle measurements: (**a**) sample 1—distilled water; (**b**) sample 1—diiodomethane; (**c**) sample 2—distilled water; (**d**) sample 2—diiodomethane; (**e**) sample 3—distilled water; (**f**) sample 3—diiodomethane; (**g**) sample 4—distilled water; and (**h**) sample 4—diiodomethane.

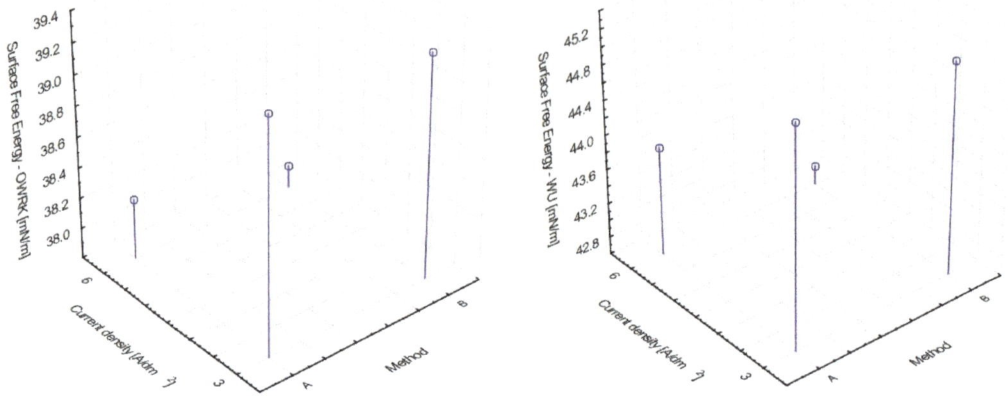

(a) (b)

Figure 10. Surface free energy plot as a function of the method of introducing IF-WS_2 NPs into the coating and current density during coating deposition: (**a**) OWRK, and (**b**) WU.

To determine the relationship between the method for obtaining coatings and contact angle measurements, Pareto charts were made for contact angles determined for distilled water (Figure 11a) and diiodomethane (Figure 12a). Pareto chart analysis (Figure 11a) revealed a linear dependence in the influence of the method of introducing IF-WS_2 NPs into the Al_2O_3 coating, the current density used to obtain Al_2O_3, the interaction between the method, and current density on the contact angle values of distilled water. In the case of diiodomethane, a significant impact on the contact angle value was observed for the method of introducing IF-WS_2 NPs into the coating and the interaction between the method and current density (Figure 12a). Figures 11b and 12b present the normal probability plots of residuals for distilled water and diiodomethane, respectively. Based on the obtained plots, it is assumed that the residuals of the statistical model follow a normal distribution, indicating that the linear model used for statistical analysis is reliable.

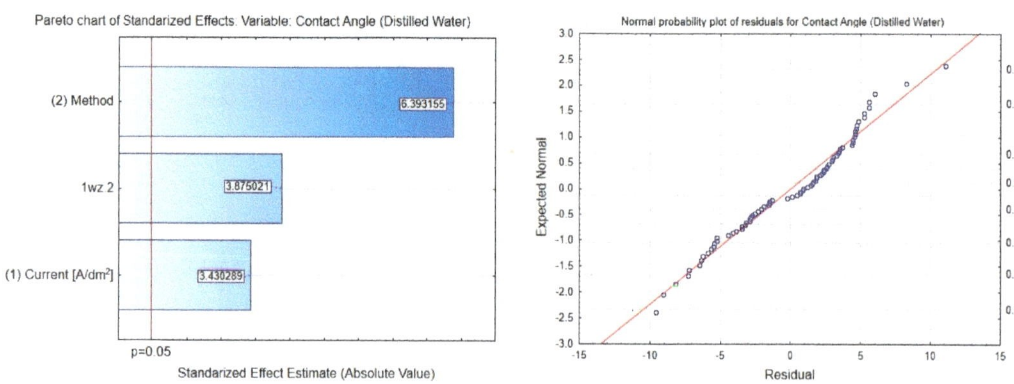

(a) (b)

Figure 11. Two-factor model with interaction for contact angle (distilled water): (**a**) the Pareto chart of standardized effect, and (**b**) the normal probability plot of residuals.

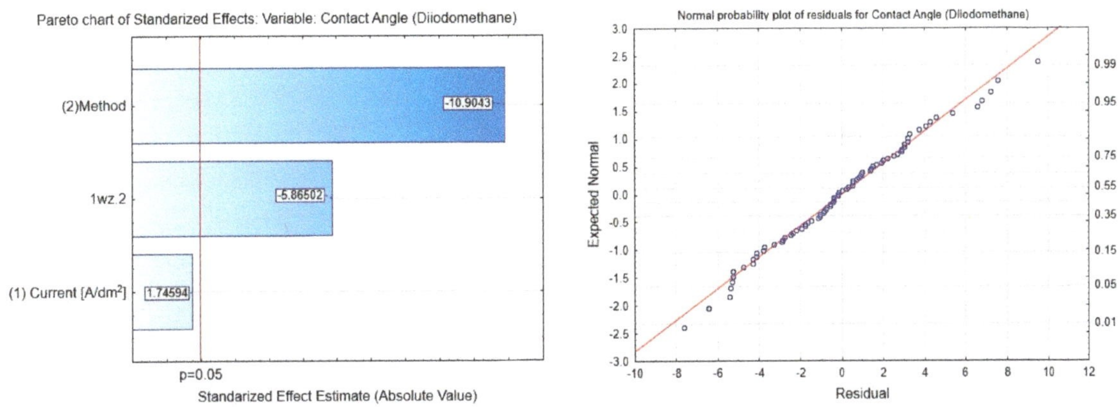

(**a**) (**b**)

Figure 12. Two-factor model with interaction for contact angle (diiodomethane): (**a**) the Pareto chart of standardized effect, and (**b**) the normal probability plot of residuals.

The research results showed that both Method A and Method B can introduce IF-WS_2 NPs into the microstructure and nanostructure of Al_2O_3 coatings. At this stage, the authors do not undertake to assess which modification method is better. The choice depends on the application of the coatings. If the goal is to reduce the friction coefficient, Method B, with larger agglomerates on the coating surface, may prove to be better. Alternatively, Method A, which hypothetically should supply NPs from the depth of the nanostructure, might be superior. This assessment will be made after tribological studies aimed at achieving the lowest possible friction coefficient and wear of the interacting element, depending on the modification method used.

4. Conclusions

To investigate the potential for better dispersion and application of IF-WS_2 NPs onto Al_2O_3 coatings, two preparation methods utilizing an ultrasonic sonicator were employed. To assess the extent of achieving the intended effect, SEM and GIXD analyses were conducted, which were correlated with surface wetting angle and surface energy studies. The main conclusions drawn from the study are as follows:

- Method A involved two stages: NP fragmentation in the sonicator and allowing samples for slow NPs deposition. Method A contributed to a more homogeneous distribution of NPs within the microstructure of Al_2O_3 coatings.
- GIXD analyses revealed the presence of a polycrystalline aluminum structure, rhombohedral phase of WS_2, and tetragonal phase of WO_3 originating from the IF-WS_2 modifier.
- The formation of WO_3 from WS_2 is likely associated with high energy during WS_2 dispersion in the sonicator.
- EDS analysis confirmed the presence of IF-WS_2 in the microstructure of Al_2O_3.
- Contact angle measurements using a non-polar liquid (diiodomethane) showed significantly lower values, which were primarily due to the liquid's properties, namely its much lower surface tension compared to water and substantially lower polarity.
- Modification of coatings using Method A resulted in surfaces with lower contact angles measured with polar liquids and higher surface free energy compared to Method B, suggesting that Method A is more effective in enhancing the hydrophilicity and surface energy of coatings. This effect may stem from the even distribution of IF-WS_2 nanoparticles in the near-surface oxide coating. In the case of Method

B, the formation of agglomerates on the oxide coating surface significantly altered surface wettability.
- Surface free energy was higher for samples anodized for 60 min at a current density of 3 A/dm^2, indicating that longer anodizing times combined with lower current density favor the creation of surfaces with higher surface energy.
- Statistical analysis demonstrated a linear relationship between the method of introducing IF-WS_2 nanoparticles into Al_2O_3 coatings and the interaction between the method and current density on the values of contact angles measured with distilled water and diiodomethane. Additionally, it showed a linear relationship between the current density applied to obtain Al_2O_3 and the contact angle values measured with distilled water.

Further research is assumed to involve the preparation of samples of appropriate sizes for tribological studies and conducting these studies, as well as examining the geometric structure of the obtained surfaces before and after wear.

Author Contributions: Conceptualization, J.K., M.N. and M.B.; methodology, J.K. and K.C.; software, J.K.; formal analysis, J.K. and M.N.; investigation, J.K., G.D., M.S. and M.N.; writing—original draft preparation, J.K. and M.N.; writing—review and editing, J.K., G.D., K.C., M.N. and M.B.; visualization, J.K.; supervision, M.B. and W.S. All authors have read and agreed to the published version of the manuscript.

Funding: This research received no external funding.

Institutional Review Board Statement: Not applicable.

Informed Consent Statement: Not applicable.

Data Availability Statement: The data presented in this study are available on request from the corresponding author.

Acknowledgments: The authors gratefully acknowledge Anna Łatkiewicz, from the Jagiellonian University, for conducting research using the scanning electron microscope.

Conflicts of Interest: The authors declare no conflicts of interest.

References

1. Kwolek, P. Corrosion Behaviour of 7075 Aluminium Alloy in Acidic Solution. *RSC Adv.* **2020**, *10*, 26078–26089. [CrossRef] [PubMed]
2. Martínez-Viademonte, M.P.; Abrahami, S.T.; Hack, T.; Burchardt, M.; Terryn, H. A Review on Anodizing of Aerospace Aluminum Alloys for Corrosion Protection. *Coatings* **2020**, *10*, 1106. [CrossRef]
3. Runge, J.M. *The Metallurgy of Anodizing Aluminum: Connecting Science to Practice*; Springer: Cham, Switzerland, 2018; ISBN 9783319721774.
4. European Aluminium. *VISION 2050 | A Vision for Strategic, Low Carbon and Competitive Aluminium*; European Aluminium: Brussels, Belgium, 2022.
5. Matijošius, T.; Padgurskas, J.; Bikulčius, G. Possibility of Phase Transformation of Al_2O_3 by a Laser: A Review. *Crystals* **2024**, *14*, 415. [CrossRef]
6. Elkilany, H.A.; Shoeib, M.A.; Abdel-Salam, O.E. Influence of Hard Anodizing on the Mechanical and Corrosion Properties of Different Aluminum Alloys. *Metallogr. Microstruct. Anal.* **2019**, *8*, 861–870. [CrossRef]
7. Kchaou, M. Friction Behavior of Anodic Oxide Layer Coating on 2017A T4 Aluminum Alloy under Severe Friction Solicitation: The Effect of Anodizing Parameters. *Eng. Technol. Appl. Sci. Res.* **2024**, *14*, 12574–12580. [CrossRef]
8. Benea, L.; Simionescu-Bogatu, N.; Chiriac, R. Electrochemically Obtained Al_2O_3 Nanoporous Layers with Increased Anticorrosive Properties of Aluminum Alloy. *J. Mater. Res. Technol.* **2022**, *17*, 2636–2647. [CrossRef]
9. Dervishi, E.; McBride, M.; Edwards, R.; Gutierrez, M.; Li, N.; Buntyn, R.; Hooks, D.E. Mechanical and Tribological Properties of Anodic Al Coatings as a Function of Anodizing Conditions. *Surf. Coat. Technol.* **2022**, *444*, 128652. [CrossRef]
10. Kwolek, P.; Obłój, A.; Kościelniak, B.; Buszta, R.; Tokarski, T.; Krupa, K.; Gradzik, A.; Nowak, W.J.; Wojnicki, M.; Motyka, M. Wear Resistance of Hard Anodic Coatings Fabricated on 5005 and 6061 Aluminum Alloys. *Arch. Civ. Mech. Eng.* **2024**, *24*, 51. [CrossRef]
11. Brudzisz, A.M.; Giziński, D.; Stępniowski, W.J. Incorporation of Ions into Nanostructured Anodic Oxides—Mechanism and Functionalities. *Molecules* **2021**, *26*, 6378. [CrossRef]

12. Kmita, T.; Bara, M. Surface Oxide Layers with an Increased Carbon Content for Applications in Oil-Less Tribological Systems. *Chem. Process Eng.-Inz. Chem. Proces.* **2012**, *33*, 479–486. [CrossRef]
13. Takaya, M.; Hashimoto, K.; Toda, Y.; Maejima, M. Novel Tribological Properties of Anodic Oxide Coating of Aluminum Impregnated with Iodine Compound. *Surf. Coat. Technol.* **2003**, *169–170*, 160–162. [CrossRef]
14. Maejima, M.; Saruwatari, K.; Takaya, M. Friction Behaviour of Anodic Oxide Film on Aluminum Impregnated with Molybdenum Sulfide Compounds. *Surf. Coat. Technol.* **2000**, *132*, 105–110. [CrossRef]
15. Posmyk, A. *Kształtowanie Właściwości Tribologicznych Warstw Wierzchnich Na Bazie Aluminium*; Wydawnictwo Politechniki Śląskiej: Gliwice, Poland, 2002.
16. Li, B.; Wang, X.; Liu, W.; Xue, Q. Tribochemistry and Antiwear Mechanism of Organic-Inorganic Nanoparticles as Lubricant Additives. *Tribol. Lett.* **2006**, *22*, 79–84. [CrossRef]
17. Wu, N.; Hu, N.; Wu, J.; Zhou, G. Tribology Properties of Synthesized Multiscale Lamellar WS_2 and Their Synergistic Effect with Anti-Wear Agent ZDDP. *Appl. Sci.* **2020**, *10*, 115. [CrossRef]
18. Gulzar, M.; Masjuki, H.H.; Kalam, M.A.; Varman, M.; Zulkifli, N.W.M.; Mufti, R.A.; Zahid, R. Tribological Performance of Nanoparticles as Lubricating Oil Additives. *J. Nanoparticle Res.* **2016**, *18*, 223. [CrossRef]
19. Rapoport, L.; Bilik, Y.; Feldman, Y.; Homyonfer, M.; Cohen, S.R.; Tenne, R. Hollow Nanoparticles of WS_2 as Potential Solid-State Lubricants. *Nature* **1997**, *387*, 791–793. [CrossRef]
20. Rapoport, L.; Leshchinsky, V.; Lapsker, I.; Volovik, Y.; Nepomnyashchy, O.; Lvovsky, M.; Popovitz-Biro, R.; Feldman, Y.; Tenne, R. Tribological Properties of WS_2 Nanoparticles under Mixed Lubrication. *Wear* **2003**, *255*, 785–793. [CrossRef]
21. Song, J.; Hu, L.; Qin, B.; Fan, H.; Zhang, Y. Fabrication and Tribological Behavior of Al_2O_3/MoS_2–$BaSO_4$ Laminated Composites Doped with in Situ Formed $BaMoO_4$. *Tribol. Int.* **2018**, *118*, 329–336. [CrossRef]
22. Erdi Korkmaz, M.; Kumar Gupta, M. Nano Lubricants in Machining and Tribology Applications: A State of the Art Review on Challenges and Future Trend. *J. Mol. Liq.* **2024**, *407*, 125261. [CrossRef]
23. Zak, A.; Feldman, Y.; Lyakhovitskaya, V.; Leitus, G.; Popovitz-Biro, R.; Wachtel, E.; Cohen, H.; Reich, S.; Tenne, R. Alkali Metal Intercalated Fullerene-like MS_2 (M = W, Mo) Nanoparticles and Their Properties. *J. Am. Chem. Soc.* **2002**, *124*, 4747–4758. [CrossRef]
24. Feldman, Y.; Zak, A.; Popovitz-Biro, R.; Tenne, R. New Reactor for Production of Tungsten Disulfide Hollow Onion-like (Inorganic Fullerene-like) Nanoparticles. *Solid State Sci.* **2000**, *2*, 663–672. [CrossRef]
25. Sawyer, W.G.; Blanchet, T.A. Lubrication of Mo, W, and Their Alloys with H_2S Gas Admixtures to Room Temperature Air. *Wear* **1999**, *225–229*, 581–586. [CrossRef]
26. Voevodin, A.A.; Zabinski, J.S. Nanocomposite and Nanostructured Tribological Materials for Space Applications. *Compos. Sci. Technol.* **2005**, *65*, 741–748. [CrossRef]
27. Sorrentino, A. Tribology of Self-Lubricating Polymer Nanocomposites. In *Self-Lubricating Composites*; Springer: Berlin/Heidelberg, Germany, 2018; pp. 105–131. [CrossRef]
28. Li, D.-l.; Jing, X.; Chen, H.; Sun, K.; Chen, W.; Huang, H.-g.; Ren, W.-j. Effects of Graphene Content in Alkaline Silicate Electrolyte on AA1060 Pure Aluminum Micro-Arc Oxidation Coating. *Int. J. Appl. Ceram. Technol.* **2024**, *21*, 1789–1801. [CrossRef]
29. Chen, L.; Liu, Z.; Shen, Q. Enhancing Tribological Performance by Anodizing Micro-Textured Surfaces with Nano-MoS_2 Coatings Prepared on Aluminum-Silicon Alloys. *Tribol. Int.* **2018**, *122*, 84–95. [CrossRef]
30. Han, X.; Thrush, S.J.; Zhang, Z.; Barber, G.C.; Qu, H. Tribological Characterization of ZnO Nanofluids as Fastener Lubricants. *Wear* **2021**, *468–469*, 203592. [CrossRef]
31. Korzekwa, J.; Gądek-Moszczak, A.; Zubko, M. Influence of the Size of Nanoparticles on the Microstructure of Oxide Coatings. *Mater. Sci.* **2018**, *53*, 709–716. [CrossRef]
32. Gorycza, T.; Dercz, G.; Prusik, K.; Pająk, L.; Łągiewska, E. Crystallite Size Determination of MgO Nanopowder from X-Ray Diffraction Patterns Registered in GIXD Technique. *Solid State Phenom.* **2010**, *163*, 177–182. [CrossRef]
33. Kyrylenko, S.; Warchoł, F.; Oleshko, O.; Husak, Y.; Kazek-Kęsik, A.; Korniienko, V.; Deineka, V.; Sowa, M.; Maciej, A.; Michalska, J.; et al. Effects of the Sources of Calcium and Phosphorus on the Structural and Functional Properties of Ceramic Coatings on Titanium Dental Implants Produced by Plasma Electrolytic Oxidation. *Mater. Sci. Eng. C* **2021**, *119*, 111607. [CrossRef]
34. Kazek-Kęsik, A.; Kalemba-Rec, I.; Simka, W. Anodization of a Medical-Grade Ti-6Al-7Nb Alloy in a $Ca(H_2PO_2)_2$-Hydroxyapatite Suspension. *Materials* **2019**, *12*, 3002. [CrossRef]
35. Owens, D.K.; Wendt, R.C. Estimation of the Surface Free Energy of Polymers. *Appl. Polym. Sci.* **1969**, *13*, 1741–1747. [CrossRef]
36. Wu, S. Calculation of Interfacial Tension in Polymer Systems. *J. Polym. Sci. Polym. Symp.* **1971**, *34*, 19–30. [CrossRef]
37. Pietraszek, J.; Dwornicka, R.; Krawczyk, M.; Kołomycki, M. The Non-Parametric Approach to the Quantification of the Uncertainty in the Design of Experiments Modelling. In Proceedings of the UNCECOMP 2017—2nd International Conference on Uncertainty Quantification in Computational Sciences and Engineering, Rhodes Island, Greece, 15–17 June 2017; pp. 598–604. [CrossRef]
38. Pietraszek, J.; Radek, N.; Goroshko, A.V.; Jana, A.; Ii, P. Challenges for the DOE Methodology Related to the Introduction of Industry 4.0. *Prod. Eng. Arch.* **2020**, *26*, 190–194. [CrossRef]
39. Dudek, A.; Lisiecka, B.; Radek, N.; Orman, Ł.J.; Pietraszek, J. Laser Surface Alloying of Sintered Stainless Steel. *Materials* **2022**, *15*, 6061. [CrossRef] [PubMed]

40. Shinde, P.A.; Jun, S.C. Review on Recent Progress in the Development of Tungsten Oxide Based Electrodes for Electrochemical Energy Storage. *ChemSusChem* **2020**, *13*, 11–38. [CrossRef]
41. Mandal, D.; Routh, P.; Nandi, A.K. A New Facile Synthesis of Tungsten Oxide from Tungsten Disulfide: Structure Dependent Supercapacitor and Negative Differential Resistance Properties. *Small* **2018**, *14*, 1702881. [CrossRef]

coatings

MDPI

Article

Determination of the Size of Complex Iron Oxide Nanoparticles Using Various Physical Experimental Methods

Airat G. Kiiamov [1,*], Anna G. Ivanova [1], Alexander N. Solodov [2], Mikhail A. Cherosov [1], Dmitrii A. Tayurskii [1] and Artur Khannanov [2]

1 Quantum Simulators Research Laboratory, Institute of Physics, Kazan Federal University, Kremlevskaya Str. 18, Kazan 420008, Russia; mcherosov@gmail.com (M.A.C.)
2 A.M. Butlerov Chemical Institute, Kazan Federal University, Kremlevskaya Str. 18, Kazan 420008, Russia; sanya.solodiv@gmail.com (A.N.S.); arthann@gmail.com (A.K.)
* Correspondence: airatphd@gmail.com

Abstract: A series of organically coated iron oxide nanoparticles obtained via the thermal decomposition of iron–oleate complexes via a "heating-up" process were investigated using the methods of transmission electron microscopy, X-ray diffraction and fine magnetometry, accompanied by elaborate mathematical analysis. The analysis of dependencies of field dependencies on the magnetization of the shape and broadening of maxima of X-ray diffraction patterns and fine refinement of transmission electron microscopy data allowed us to demonstrate that all of the samples under consideration had a tripartite structure: (i) the magnetic crystalline core of iron oxide, (ii) the paramagnetic stratum of amorphous iron oxide and (iii) the organic coater. The new approach toward synthesis for organic coated iron oxide shows that it could be applied to the preparation of magnetic nanoparticles with different and controlled magnetic properties and sizes depending on necessary applications, especially biomedical.

Keywords: magnetic nanoparticles; X-ray diffraction; superparamagnetism; Langevin approach

1. Introduction

To date, magnetic nanoparticles (MNPs) have been considered as the most widely used and studied nanomaterials [1,2]. This is because MNPs offer many advantages over other nanoparticles: (i) the adjustable sizes and morphology within limits of from approximately three to several hundred nanometers, (ii) the tailor-made surface coating, which can be adapted in a way so that the particles can selectively bind to a defined biologic entity [3], (iii) low toxicity [4] and (iv) the ability to convert magnetic field energy into thermal and/or mechanical energy [5,6]. The combination of these properties opens up great possibilities for using MNPs in various applications such as multi-terabit storage data [7], catalysis [8], targeted drug delivery [1], magnetic hyperthermia [9–11], special high-contrast agents for magnetic resonance imaging (MRI) [12], biosensors and bioseparation [13], thermoablation and the development of different smart materials [14–18]. The properties of MNPs heavily depend on their size and shape. For example, for the use of MNPs as a contrast agent in MRI, particle size plays a key role; thus, large-size MNPs are used as T2 contrast agents, and smaller particles are used as T1 contrast agents [19]. Magnetic hyperthermia, one of the most important applications of MNPs, is considered as a promising cancer treatment due to its low rate of side effects. In magnetic hyperthermia, heat generation by MNPs is based on their capability of dissipating magnetic energy to heat [20,21]. This effect is influenced by parameters such as size and concentration of MNPs in the fluid, as well as magnetic parameters of MNPs such as saturation magnetization (Ms) and magnetic anisotropy constant. Among these parameters, the particle size plays a key role, because it specifies the magnetic behavior type and scope of application of MNPs.

Often, the size of a spherical MNP is understood as its diameter, determined by images obtained using bright-field transmission electron microscopy [22].

The size of a nanoparticle determined using TEM is the physical diameter of MNPs, which does not always reflect its magnetic properties. It can be expected that MNPs of the same size should all have the same magnetic moment; nevertheless, due to crystal defects, the magnetic polydispersity is likely to be greater than expected from the geometric size polydispersity [23,24]. Such effects include a "magnetically dead" surface layer and internal defects of the crystal structure [25,26].

Biomedicine is a promising field of application of MNPs, and so the size of nanoparticles plays a key role. This is important to note as the biodistribution, and ultimately biological responses, for example, of the material can be correlated sizes [27]. To this end, the ability to characterize sizes of particles is the most widely accepted critical quality attribute [28]. However, the results of the sizes of MNPs obtained via the only approach from the TEM, for instance, are generally not applicable to the comprehensive characterizing of all of the properties of the samples, especially in the case of multilayer particles. So, in the case of particles made of strong magnetic material cores shelled by the nonmagnetic coating, the TEM approaches are only able to measure the outward diameter of the particle, whereas the magnetic properties of the particles ought to depend on the size of the magnetic core. Hence, in this case, usage based on TEM measurements of particle sizes for the analysis of the magnetic properties provides the systematic miscalculation.

So, in the present study, we use three experimental methods to measure the sizes: (i) TEM to measure the full outer size, (ii) fine analysis of X-ray diffraction data in order to estimate the crystalline size and (iii) utilizing of magnetization field dependencies to obtain the size of the magnetic part of the particles. And, running ahead, it could be said that all of the sizes obtained via each method for each sample are presented in Table 1.

Table 1. The particle sizes obtained via different methods.

<table>
<tr><th>Sample ID Number</th><th>Obtained via TEM, nm</th><th>Obtained via XRD, nm</th><th>Obtained from the Field Dependencies of Magnetization, nm</th></tr>
<tr><td rowspan="4">1</td><td rowspan="4">4.82 ± 1.10</td><td>4.7</td><td rowspan="4">3.50</td></tr>
<tr><td>4.3</td></tr>
<tr><td>4.5</td></tr>
<tr><td>4.7</td></tr>
<tr><td rowspan="3">2</td><td rowspan="3">5.34 ± 1.13</td><td>5.6</td><td rowspan="3">3.33</td></tr>
<tr><td>4.0</td></tr>
<tr><td>5.3</td></tr>
<tr><td rowspan="3">3</td><td rowspan="3">6.90 ± 0.90</td><td>6.6</td><td rowspan="3">1.80</td></tr>
<tr><td>7.4</td></tr>
<tr><td>6.5</td></tr>
<tr><td rowspan="2">4</td><td rowspan="2">7.15 ± 1.62</td><td>6.8</td><td rowspan="2">4.50</td></tr>
<tr><td>6.6</td></tr>
<tr><td rowspan="3">5</td><td rowspan="3">8.02 ± 1.30</td><td>4.6</td><td rowspan="3">4.33</td></tr>
<tr><td>7.1</td></tr>
<tr><td>4.9</td></tr>
<tr><td rowspan="4">6</td><td rowspan="4">8.43 ± 0.95</td><td>4.9</td><td rowspan="4">4.25</td></tr>
<tr><td>8.0</td></tr>
<tr><td>7.6</td></tr>
<tr><td>4.8</td></tr>
</table>

Table 1. *Cont.*

Sample ID Number	Obtained via TEM, nm	Obtained via XRD, nm	Obtained from the Field Dependencies of Magnetization, nm
7	9.44 ± 1.64	5.4	4.00
		7.0	
		5.3	
8	9.78 ± 2.00	6.3	3.30
		8.2	
		6.5	
9	10.26 ± 1.88	6.3	3.33
		10.1	
		7.4	
		8.0	
10	10.48 ± 1.86	7.9	3.33
		10.3	
		10.0	

Nanoparticles with low geometric size polydispersity can be prepared via the thermal decomposition of organometallic precursors [29–31]. Obtaining uniformity of the size of MNPs in this synthesis method is achieved by separating the processes of nucleation and growth. A brief nucleation event yields initial particles that subsequently all grow at the same rate under conditions where no new nuclei are formed. However, recent work has uncovered that the thermal decomposition routes may actually proceed via the formation of a nonmagnetic Wüstite phase, and that the formation of the magnetic magnetite/maghemite phases proceed afterward due to oxidation upon exposure to atmospheric oxygen [32,33]. Therefore, the geometric size and magnetic size for MNPs may differ.

In this paper, we present three approaches toward the synthesis of MNPs based on iron oxides via thermal decomposition: (i) the classical approach of obtaining MNPs via thermal decomposition, proposed by Park [31], (ii) using seed crystals and (iii) the joint thermal decomposition of iron (II) and (III) precursors. We also compare the dimensions obtained via transmission electron microscopy, magnetometry and XRD.

2. Materials and Methods

Materials.

Sodium hydroxide, n-hexane and ethanol were supplied by JSC "TatKhim Product" LLC (Kazan, Republic of Tatarstan, Russia); iron(III) chloride hexahydrate ($FeCl_3{\cdot}6H_2O$, 99%), iron(II) chloride tetrahydrate ($FeCl_2{\cdot}4H_2O$, 99%), oleic acid ($C_{18}H_{34}O_2$, 90%) and 1-octadecene ($C_{18}H_{36}$, 90%) were obtained from Sigma-Aldrich (Burlington, MA, USA).

Methods.

Synthesis of iron–oleate complex. The metal–oleate complex was prepared by reacting metal chlorides and sodium oleate. Deionized water (60 mL) was mixed with ethanol (80 mL) and divided into two equal portions. Sodium hydroxide (9.6 g) was added to the first portion (the first solution), and $FeCl_3{\cdot}6H_2O$ (21.6 g) for the iron(III)–oleate complex or $FeCl_2{\cdot}4H_2O$ (23.9 g) for the iron(II)–oleate complex were added to the second portion (the second solution). Both solutions were successively added to a solution of oleic acid (67.8 g) in 280 mL of n-hexane. The mixture was magnetically stirred and refluxed for 4 h at 70 °C. After the reflux, the flask was removed from the oil bath and cooled to room temperature. Next, the upper organic layer containing the iron–oleate complex was washed three times with 100 mL distilled water in a separatory funnel. After washing, hexane was evaporated off. The product yield was 98%.

Synthesis of iron oxide nanocrystals. (i) The MNPs (Nos. 1, 3, 5 and 6) were synthesized according to the Park method [31,34]. We used the same synthesis procedure and materials as declared in our previous study on a similar kind of sample [34]. In a typical synthesis of MNPs (No. 1), 3.6 g of iron(III)–oleate complex was dissolved in 25 g of 1-octadecene at room temperature. The reaction mixture was heated to 320 °C with a constant heating rate of 3.3 °C min^{-1}, and then kept at that temperature for 30 min. The resulting solution containing the nanocrystals was then cooled to room temperature, and 500 mL of ethanol was added to the solution to precipitate the nanocrystals. The nanocrystals were separated via centrifugation.

Sample Nos. 3, 5 and 6 were obtained via the method described above. For particle No. 3, the solution contained 3.6 g of iron(III)–oleate complex and 35 g of 1-octadecene; for particle No. 5, the solution contained 3.6 g of iron(III)–oleate complex, 0.57 g of oleic acid and 25 g of 1-octadecene; for particle No. 6, the solution contained 3.6 g of iron(III)–oleate complex, 0.57 g of oleic acid and 15 g of 1-octadecene.

(ii) For the synthesis of MNP Nos. 2, 4 and 9, the thermal decomposition method was used, as in the Park method, but seed crystals were additionally used. For particle No. 2, the solution contained 3.6 g of iron(III)–oleate complex, 0.1 g of MNPs (No. 1) and 25 g of 1-octadecene; for particle No. 4, the solution contained 3.6 g of iron(III)–oleate complex, 0.58 g of oleic acid, 0.5 g of MNPs (No. 1) and 25 g of 1-octadecene; and for particle No. 9, the solution contained 3.6 g of iron(III)–oleate complex, 0.57 g of oleic acid, 0.1 g of MNPs (No. 1) and 25 g of 1-octadecene.

For the synthesis of particle Nos. 7, 8 and 10, the thermal decomposition method was used, as in the Park method, but the addition of iron(II)–oleate complex was additionally used. For particle No. 7, the solution contained 2.4 g of iron(III)–oleate complex, 1.2 g of iron(II)–oleate complex, 0.57 g of oleic acid and 25 g of 1-octadecene; for particle No. 8, the solution contained 2.4 g of iron(III)–oleate complex, 1.2 g of iron(II)–oleate complex and 25 g of 1-octadecene; for particle No. 10, the solution contained 2.4 g of iron(III)–oleate complex, 1.2 g of iron(II)–oleate complex, 1.20 g of oleic acid and 25 g of 1-octadecene.

We used the transmission electron microscope Hitachi HT7700 Excellence to provide transmission electron microscopy imaging (with an accelerated voltage value of 100 kV).

The powder XRD (X-ray diffraction) measurements were taken using a Bruker D8 Advance diffractometer (Cu Kα radiation λ = 1.5418 Å) with symmetric measurement geometry; the measurement time was 4 s per step and the diffraction angle range was of 7°–100°. We used the Scherrer equation to estimate the size of the crystalline iron oxide particles [34].

The magnetization measurements were taken using a commercial PPMS-9 machine and a commercial vibrational magnetometer at a temperature of 300 K with a magnetic field value up to 9 T. The obtained magnetization curves were used to determine the size of magnetic particles via the Langevin function. The details of the fitting procedure and data analysis are given below in the next section.

We conducted the Fourier transform infrared spectroscopy measurements with 4 cm^{-1} step length and wave numbers ranging from 400 to 4000 cm^{-1} using an FT-801 spectrometer with attenuated total reflection.

3. Results

We obtained 10 types of MNPs via the thermal decomposition (Park's method) of iron–oleate complexes in an oleic acid/octadecene solution at a temperature of 320 °C. In addition to the thermal decomposition of iron–oleate complexes to the corresponding oxides, the resulting oxides were partially reduced, resulting in magnetite or iron oxide nanoparticles (II/III). In this regard, the presence of both iron(II) and iron(III) in the system should facilitate the formation of nanoparticles and improve their magnetic properties.

TEM images of the MNP samples and the corresponding particle size distribution obtained via the statistical analysis of over ~3500 particles are shown in Figure 1. The nanocrystals exhibit a well-defined spherical morphology, are faceted and are nearly monodispersed in size and shape. The particle size of the spherical samples increases from sample No. 1 to No. 10; the average diameter of the samples is shown in Figure 1.

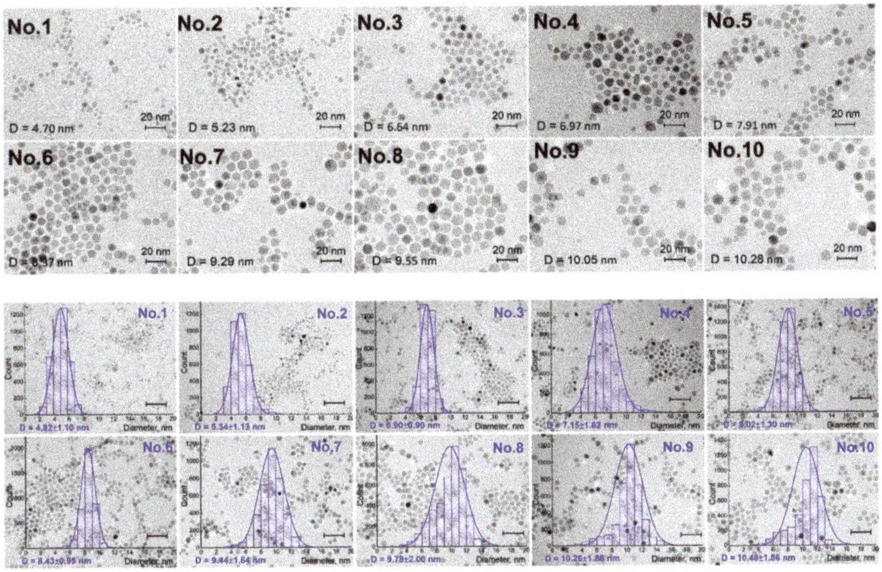

Figure 1. TEM images (**top section**) and the corresponding particle size distribution (**bottom section**) obtained via statistical analysis of over ~3500 particles. In the (**bottom section**), all scale bars are 40 nm.

The FT-IR spectra of the MNPs are presented in Figure 2. A typical MNP's spectrum shows a peak at 554 cm^{-1} which corresponds to the Fe-O bond valence vibration range. The absorbance bands of the organic surfactant on the surface of the particles were clearly observed at higher wavenumbers.

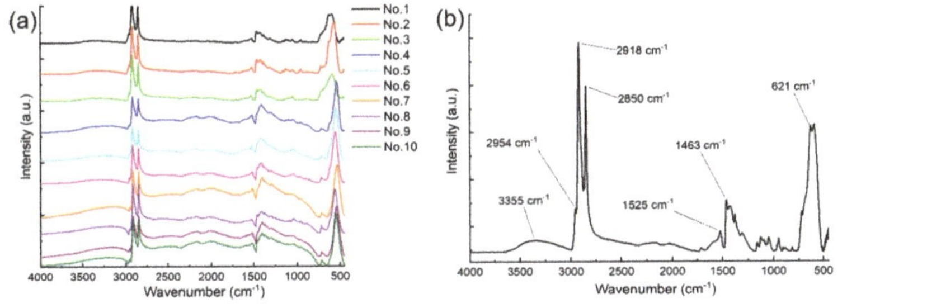

Figure 2. (**a**) FT-IR spectra of the synthesized MNPs (Nos. 1–10). (**b**) FT-IR spectra of MNP No. 1.

The iron oxide nanocrystals were characterized by X-ray powder diffraction (XRD), and the XRD patterns (Figure 3) of the nanocrystals were assigned to the (220), (311), (400), (422), (511) and (440) reflections of the inverse spinel structure iron oxide. In all of the XRD patterns, a clear line broadening can be observed, which results from size effects. This peak broadening can be utilized to estimate the mean crystallite size and the lattice strain in particles due to crystal imperfection and distortion. As the particle size diminishes, the XRD peaks become wider.

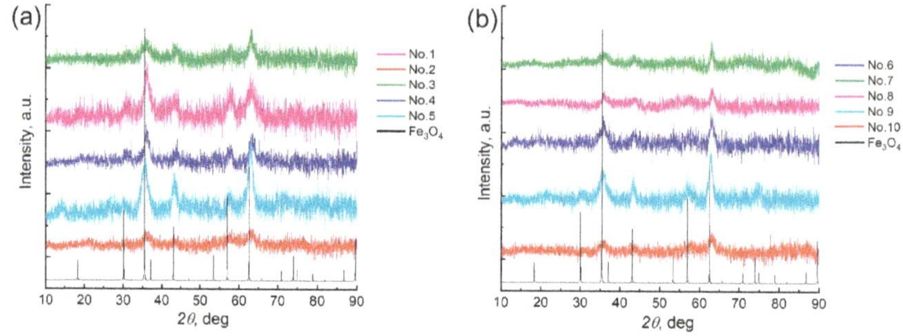

Figure 3. Powder XRD patterns of the synthesized MNPs: (**a**) Nos. 1–5, (**b**) Nos. 6–10.

The Scherer formula approach was used to determine the crystallite size and estimate the shape of iron oxide nanoparticles from the corresponding X-ray diffraction patterns. In accordance with the Scherrer formula, the particle size D is inversely proportional to the peak width β:

$$D = K\frac{\lambda}{\beta\ cos\theta},$$

where K means the shape factor (we used $K = 0.9$), θ is a half of the diffraction angle 2θ corresponding to the diffraction maximum under consideration and λ means the used X-ray radiation wavelength (1.5418 Å). The angular positions 2θ and full width at half maximum β for all of the diffraction maxima of each sample were determined using Wolfram Mathematica 6.0 software, and each diffraction maximum was described using the Lorentzian line shape.

For each sample, the particle sizes along different crystallographic directions were estimated using an analysis of the diffraction maxima from the corresponding set of crystallographic planes. As seen via the comparison of particle sizes along various spatial dimensions, the shapes of particles were characterized as being spherical or non-spherical. In order to formularize the criteria for the sphericity of a particle, we propose to consider a nanoparticle as being spherical if its average size, obtained as the arithmetic mean of all measured crystallite sizes along different crystallographic directions, differs from the crystallite size along any measured crystallographic direction by no more than 10 percent.

Thus, the shapes of iron oxide nanoparticles of only two samples (3 and 4) were estimated as being spherical. The iron oxide nanoparticles of the sample numbers 5, 7 and 9 are relatively elongated along the (511) or (440) crystallographic directions, which allows us to make assumptions about the cigar shape of the crystallites in the case of these samples.

The field dependencies of the magnetization $M(H)$ of the samples under consideration measured via sample temperature fine stabilization with the temperature T of 300 K are shown in Figure 4. The Langevin function [L(x) = coth(x) − 1/x] describing the degree of alignment of a magnetic dipole in an external field H was used to approximate the measured field dependencies of magnetization.

$$M(H) = \int_0^{\infty} p(D_m; \mu; \sigma)[M_s L(x) + \chi H]dD_m$$

where

$$x = \frac{\pi\mu_0 M_s D_m^3 H}{6k_B T}$$

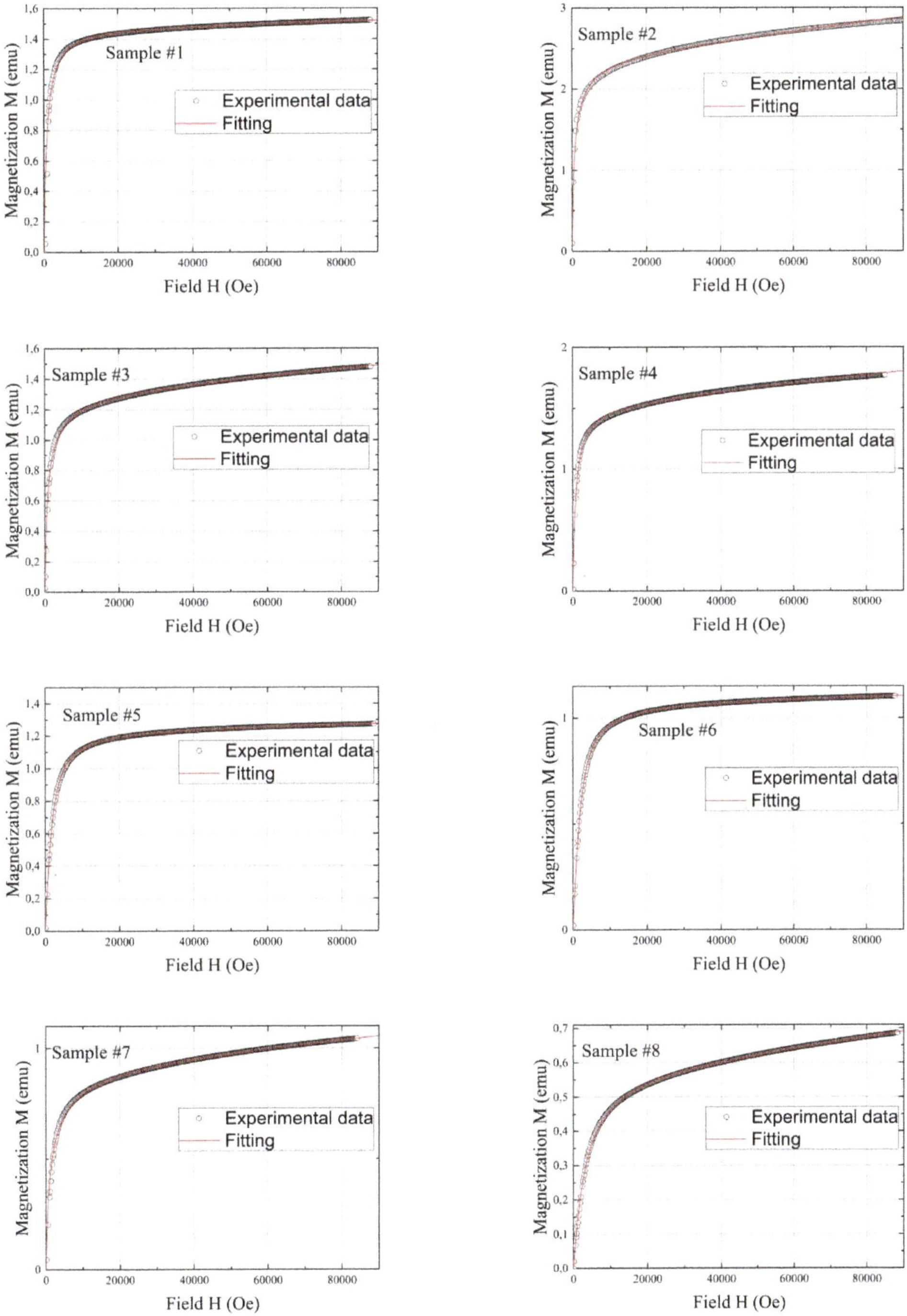

Figure 4. *Cont.*

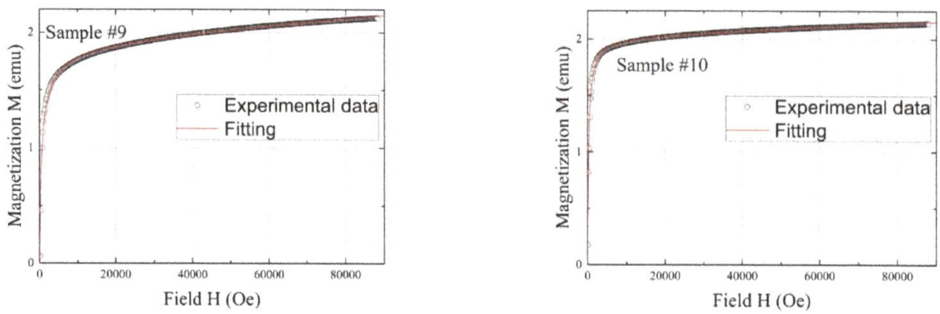

Figure 4. Field dependencies of magnetization M (emu) of MNPs (open black circles) and its approximation (thin red line) within the Langevin function weighted using a log-normal size distribution (see details in the text).

The Boltzmann constant is written as k_B, whereas M_s is the saturated magnetization and μ_0 is the magnetic constant. The function was weighted using a log-normal size distribution $p(D_m;\sigma) = \frac{1}{\sqrt{2\pi}\sigma D_m} exp\left[-\frac{(ln^2(D_m/D_M)}{2\sigma^2}\right]$ with a dispersion of σ in order to take into account the non-uniformity of particle sizes D_m, as has been proposed in [35]. Finally, the simple paramagnetic term χH (χ is a paramagnetic susceptibility) was also used in order to take into account the contribution of amorphous iron oxide which surrounded the crystalline superparamagnetic core. As a result of the analysis of magnetic data, the size of each sample of magnetic core D_M was estimated. It deserves to be noted that we used the Wolfram Mathematica 6.0 software for approximation.

4. Discussion

The estimated sizes of particles, crystalline parts and magnetic cores are collected in Table 1. The full particle sizes were obtained using TEM methods, while the crystalline size was found as a result of the approximation of XRD data. It should be reinstated that by means of X-ray diffraction analysis methods, we can only estimate the size of the crystalline part of the particles. So, Table 1 compares three differently estimated sizes of the particles: (i) the real particle size measured using TEM approaches, (ii) the crystalline particle size obtained via means of XRD and (iii) the magnetic core size estimated from the fine analysis of field dependencies of magnetization.

The results presented above confirm that the synthesis of shelled iron oxide magnetic nanoparticles was performed well using ten different procedures. The good agreement and consistency between the particle size estimations obtained via three different methods are clearly visible.

It was proven that particles have a multilayer structure with crystalline cores of iron oxide coated by the organic shells. These iron oxide cores are stand-alone magnets and, in unison, act as a superparamagnetic conglomerate. Nevertheless, we note the kind of minor divergence of the size of data obtained within XRD and magnetometry approaches. We argue that this disagreement is not a result of experimental error; we believe that it could be due to the fact that the upper-bound part of an iron oxide core gives no contribution to the total magnetic moment, since the surface spins are not exchange interactions coupled with the spins in the volume of the magnetic cores of the nanoparticles.

So, it could be confidently argued that the particles have a tripartite structure: (i) the magnetic crystalline core of iron oxide, (ii) the paramagnetic stratum of amorphous iron oxide and (iii) the organic coater.

5. Conclusions

We conclude that it has been proven that directly measured TEM particle sizes are not equal to the crystalline size obtained via XRD approaches nor the effective magnetic particle size. An effective magnetic particle size means that the magnetic behavior of the particles resembles a behavior of a superparamagnetic assembly with the particles of the same size as effective magnetic particle size. However, the difference obtained between the TEM particle sizes and the data obtained via XRD evidence the fact that the crystalline particles are covered with a non-crystalline organic oleate shell.

It could be concluded that the presented synthesis procedures allow for preparing MNPs with different and controlled magnetic properties and sizes depending on necessary applications, especially biomedical, as, as considered in the present study, nanoparticles are coated with a bio-friendly shell. Nevertheless, it should be noted that further possible medical application of MNPs should be prefaced with reliable and strict medical verification.

Author Contributions: Conceptualization, A.K.; Methodology, A.G.K. and A.N.S.; Software, A.G.K.; Formal analysis, A.G.K. and A.G.I.; Investigation, A.G.K. and M.A.C.; Resources, A.N.S. and A.K.; Data curation, A.G.K., A.G.I. and A.N.S.; Writing—original draft, A.G.K. and A.N.S.; Writing—review & editing, A.G.K.; Supervision, D.A.T.; Project administration, D.A.T. and A.K.; Funding acquisition, A.K. All authors have read and agreed to the published version of the manuscript.

Funding: The study was supported by a grant from the Russian Science Foundation, No. 22-73-10036 (https://rscf.ru/project/22-73-10036/ accessed on 11 September 2023). This work was supported by the Kazan Federal University Strategic Academic Leadership Program (PRIORITY-2030).

Institutional Review Board Statement: Not applicable.

Informed Consent Statement: Not applicable.

Data Availability Statement: Not applicable.

Conflicts of Interest: The authors declare no conflict of interest.

References

1. Laurent, S.; Forge, D.; Port, M.; Roch, A.; Robic, C.; Vander Elst, L.; Muller, R.N. Magnetic iron oxide nanoparticles: Synthesis, stabilization, vectorization, physicochemical characterizations, and biological applications. *Chem. Rev.* **2008**, *108*, 2064–2110. [CrossRef] [PubMed]
2. Sezer, N.; Arı, İ.; Bicer, Y.; Koc, M. Superparamagnetic nanoarchitectures: Multimodal functionalities and applications. *J. Magn. Magn. Mater.* **2021**, *538*, 168300. [CrossRef]
3. Veiseh, O.; Gunn, J.W.; Zhang, M. Design and fabrication of magnetic nanoparticles for targeted drug delivery and imaging. *Adv. Drug Deliv. Rev.* **2010**, *62*, 284–304. [CrossRef] [PubMed]
4. Mahmoudi, M.; Hofmann, H.; Rothen-Rutishauser, B.; Petri-Fink, A. Assessing the in vitro and in vivo toxicity of superparamagnetic iron oxide nanoparticles. *Chem. Rev.* **2012**, *112*, 2323–2338. [CrossRef]
5. Nandhini, G.; Shobana, M. Role of ferrite nanoparticles in hyperthermia applications. *J. Magn. Magn. Mater.* **2022**, *552*, 169236. [CrossRef]
6. Zhang, C.; Li, X.; Jiang, L.; Tang, D.; Xu, H.; Zhao, P.; Fu, J.; Zhou, Q.; Chen, Y. 3D printing of functional magnetic materials: From design to applications. *Adv. Funct. Mater.* **2021**, *31*, 2102777. [CrossRef]
7. Weller, D.; Moser, A. Thermal effect limits in ultrahigh-density magnetic recording. *IEEE Trans. Magn.* **1999**, *35*, 4423–4439. [CrossRef]
8. Weiss, W.; Ranke, W. Surface chemistry and catalysis on well-defined epitaxial iron-oxide layers. *Prog. Surf. Sci.* **2002**, *70*, 1–151. [CrossRef]
9. Rytov, R.; Bautin, V.; Usov, N. Towards optimal thermal distribution in magnetic hyperthermia. *Sci. Rep.* **2022**, *12*, 3023. [CrossRef]
10. Usov, N.A.; Gubanova, E.M. Application of magnetosomes in magnetic hyperthermia. *Nanomaterials* **2020**, *10*, 1320. [CrossRef]
11. Tong, S.; Quinto, C.A.; Zhang, L.; Mohindra, P.; Bao, G. Size-dependent heating of magnetic iron oxide nanoparticles. *ACS Nano* **2017**, *11*, 6808–6816. [CrossRef] [PubMed]
12. Na, H.B.; Song, I.C.; Hyeon, T. Inorganic nanoparticles for MRI contrast agents. *Adv. Mater.* **2009**, *21*, 2133–2148. [CrossRef]
13. Khizar, S.; Ahmad, N.M.; Zine, N.; Jaffrezic-Renault, N.; Errachid-el-salhi, A.; Elaissari, A. Magnetic nanoparticles: From synthesis to theranostic applications. *ACS Appl. Nano Mater.* **2021**, *4*, 4284–4306. [CrossRef]
14. Xu, Z.; Xu, Q. Collective behaviors of magnetic microparticle swarms: From dexterous tentacles to reconfigurable carpets. *ACS Nano* **2022**, *16*, 13728–13739. [CrossRef]
15. Khamidullin, T.; Lounev, I.; Solodov, A.; Vakhitov, I.; Hashemi, S.A.; Dimiev, A.M. Graphene Oxide–Epoxy Composites with Induced Anisotropy of Electrical Properties. *J. Phys. Chem. C* **2021**, *125*, 26823–26831. [CrossRef]

16. Solodov, A.N.; Balkaev, D.; Amirov, R.; Gataullina, R.; Nurtdinova, L.; Yusupov, R.; Kharintsev, S.S.; Dimiev, A.M. Polymer Composites with Magnetically Tunable Optical Anisotropy. *ACS Appl. Polym. Mater.* **2023**, *5*, 6338–6345. [CrossRef]
17. Solodov, A.; Neklyudov, V.V.; Shayimova, J.; Amirov, R.R.; Dimiev, A.M. Magneto-optical properties of the magnetite-graphene oxide composites in organic solvents. *ACS Appl. Mater. Interfaces* **2018**, *10*, 40024–40031. [CrossRef]
18. Solodov, A.N.; Balkaev, D.A.; Shayimova, J.R.; Vakhitov, I.R.; Gataullina, R.M.; Sukhov, A.V.; Burilova, E.A.; Amirova, L.M.; Zhuravleva, Y.I.; Amirov, R.R. Tribological properties of an epoxy polymer containing a magnetically oriented graphene oxide/iron oxide nanoparticle composite. *Diam. Relat. Mater.* **2023**, *138*, 110211. [CrossRef]
19. Zhou, Z.; Yang, L.; Gao, J.; Chen, X. Structure–relaxivity relationships of magnetic nanoparticles for magnetic resonance imaging. *Adv. Mater.* **2019**, *31*, 1804567. [CrossRef]
20. Tang, Y.; Flesch, R.C.; Jin, T.; Gao, Y.; He, M. Effect of nanoparticle shape on therapeutic temperature distribution during magnetic hyperthermia. *J. Phys. D Appl. Phys.* **2021**, *54*, 165401. [CrossRef]
21. Nemati, Z.; Alonso, J.; Rodrigo, I.; Das, R.; Garaio, E.; García, J.Á.; Orue, I.; Phan, M.-H.; Srikanth, H. Improving the heating efficiency of iron oxide nanoparticles by tuning their shape and size. *J. Phys. Chem. C* **2018**, *122*, 2367–2381. [CrossRef]
22. Lim, J.; Yeap, S.P.; Che, H.X.; Low, S.C. Characterization of magnetic nanoparticle by dynamic light scattering. *Nanoscale Res. Lett.* **2013**, *8*, 381. [CrossRef]
23. Unni, M.; Uhl, A.M.; Savliwala, S.; Savitzky, B.H.; Dhavalikar, R.; Garraud, N.; Arnold, D.P.; Kourkoutis, L.F.; Andrew, J.S.; Rinaldi, C. Thermal decomposition synthesis of iron oxide nanoparticles with diminished magnetic dead layer by controlled addition of oxygen. *ACS Nano* **2017**, *11*, 2284–2303. [CrossRef] [PubMed]
24. Luigjes, B.; Woudenberg, S.M.; de Groot, R.; Meeldijk, J.D.; Torres Galvis, H.M.; de Jong, K.P.; Philipse, A.P.; Erné, B.H. Diverging geometric and magnetic size distributions of iron oxide nanocrystals. *J. Phys. Chem. C* **2011**, *115*, 14598–14605. [CrossRef]
25. Kodama, R. Magnetic nanoparticles. *J. Magn. Magn. Mater.* **1999**, *200*, 359–372. [CrossRef]
26. Roca, A.; Morales, M.; O'Grady, K.; Serna, C. Structural and magnetic properties of uniform magnetite nanoparticles prepared by high temperature decomposition of organic precursors. *Nanotechnology* **2006**, *17*, 2783. [CrossRef]
27. Hassellöv, M.; Readman, J.W.; Ranville, J.F.; Tiede, K. Nanoparticle analysis and characterization methodologies in environmental risk assessment of engineered nanoparticles. *Ecotoxicology* **2008**, *17*, 344–361. [CrossRef]
28. Maguire, C.M.; Rösslein, M.; Wick, P.; Prina-Mello, A. Characterisation of particles in solution–a perspective on light scattering and comparative technologies. *Sci. Technol. Adv. Mater.* **2018**, *19*, 732–745. [CrossRef] [PubMed]
29. Park, J.; Lee, E.; Hwang, N.M.; Kang, M.; Kim, S.C.; Hwang, Y.; Park, J.G.; Noh, H.J.; Kim, J.Y.; Park, J.H. One-nanometer-scale size-controlled synthesis of monodisperse magnetic Iron oxide nanoparticles. *Angew. Chem.* **2005**, *117*, 2932–2937. [CrossRef]
30. Sun, S.; Zeng, H.; Robinson, D.B.; Raoux, S.; Rice, P.M.; Wang, S.X.; Li, G. Monodisperse mfe2o4 (m = fe, co, mn) nanoparticles. *J. Am. Chem. Soc.* **2004**, *126*, 273–279. [CrossRef]
31. Park, J.; An, K.; Hwang, Y.; Park, J.-G.; Noh, H.-J.; Kim, J.-Y.; Park, J.-H.; Hwang, N.-M.; Hyeon, T. Ultra-large-scale syntheses of monodisperse nanocrystals. *Nat. Mater.* **2004**, *3*, 891–895. [CrossRef]
32. Chalasani, R.; Vasudevan, S. Form, content, and magnetism in iron oxide nanocrystals. *J. Phys. Chem. C* **2011**, *115*, 18088–18093. [CrossRef]
33. Casula, M.F.; Jun, Y.-W.; Zaziski, D.J.; Chan, E.M.; Corrias, A.; Alivisatos, A.P. The concept of delayed nucleation in nanocrystal growth demonstrated for the case of iron oxide nanodisks. *J. Am. Chem. Soc.* **2006**, *128*, 1675–1682. [CrossRef] [PubMed]
34. Solodov, A.N.; Shayimova, J.R.; Burilova, E.A.; Shurtakova, D.V.; Zhuravleva, Y.I.; Cherosov, M.A.; Tian, Y.; Kiiamov, A.G.; Amirov, R.R. Understanding the Nucleation and Growth of Iron Oxide Nanoparticle Formation by a "Heating-Up" Process: An NMR Relaxation Study. *J. Phys. Chem. C* **2021**, *125*, 20980–20992. [CrossRef]
35. Chen, D.-X.; Sanchez, A.; Taboada, E.; Roig, A.; Sun, N.; Gu, H.-C. Size determination of superparamagnetic nanoparticles from magnetization curve. *J. Appl. Phys.* **2009**, *105*, 083924. [CrossRef]

coatings

Article

Evolution of the Growth of a Micro-Nano Crystalline Diamond Film on an Axial Carbide Tool Model in Microwave Plasma

Evgeny E. Ashkinazi [1], Sergey V. Fedorov [2,*], Artem K. Martyanov [1], Vadim S. Sedov [1], Alexey F. Popovich [1,3], Andrey P. Bolshakov [1], Dmitry N. Sovyk [1], Stanislav G. Ryzhkov [1], Andrey A. Khomich [3], Evgeny V. Zavedeev [1], Artem P. Litvinov [4], Vladimir Y. Yurov [1], Sergey N. Grigoriev [2] and Vitaly I. Konov [1]

1 Department of Light-Induced Surface Phenomena, Prokhorov General Physics Institute of the Russian Academy of Sciences, Vavilov Str. 38, 119991 Moscow, Russia; jane50@list.ru (E.E.A.); art.martyanov@gmail.com (A.K.M.); sedovvadim@yandex.ru (V.S.S.); lex78@mail.ru (A.F.P.); bolshak3@yandex.ru (A.P.B.); sovyk@inbox.ru (D.N.S.); high-low@yandex.ru (S.G.R.); dodeskoden@gmail.com (E.V.Z.); yurov6591@gmail.com (V.Y.Y.); vitaly.konov@gmail.com (V.I.K.)

2 Department of High-Efficiency Machining Technologies, Moscow State University of Technology STANKIN, Vadkovskiy per. 3A, 127055 Moscow, Russia; s.grigoriev@stankin.ru

3 Semiconductor Growth Laboratory MOCVD, Fryazinsky Branch of the V.A. Kotelnikov Institute of Radio Engineering and Electronics of the Russian Academy of Sciences, 1 Vvedensky Square, 141190 Fryazino, Russia; antares-610@yandex.ru

4 Department of Technological Systems, Moscow Polytechnic University, B. Semenovskaya Str. 38, 107023 Moscow, Russia; mrartem.litvinov35801@mail.ru

* Correspondence: sv.fedorov@stankin.ru; Tel.: +7-916-290-26-07

Abstract: Conformal multilayer micro-nanocrystalline diamond coatings were grown on substrates of a hard alloy with 9% Co with a high aspect number in microwave plasma from gas mixtures CH4/H2 and CH4/H2/N2. The substrates were of a cylindrical axial tool model with a size ratio: d = 12 mm to l = 75 mm. An original tool holder made of molybdenum, in the form of a sector of the excessive ring with the axis of the hole parallel to the central conductive platform, protects part of the substrate from heating due to the edge effect of plasma. The uniformity of heating of the growth part, which is located inside the excessive ring, is calculated using mathematical modeling and is provided by rotation at a speed of at least 12 rpm, which ensures the uniformity of the coating. The average grain size of the nanocrystalline film measured along the cylinder forming was 41 nm.

Keywords: diamond; tungsten carbide axial tool model; multi-layered films; chemical vapor deposition; microwave plasma

1. Introduction

Diamond coatings on WC-Co hard alloy tools, due to their high hardness and thermal conductivity, low coefficient of friction, strength, and wear resistance, play an essential role in mastering the production of promising materials [1–4]. The development of new methods for applying polycrystalline diamond coatings (DC) directly to the working surface of base materials enables the manufacture of axial tools using the best properties of diamond. Axial tools (cutters, drills, mills) with chemical vapors deposition (CVD) diamond coating are mainly used for processing carbon-fiber-reinforced plastic (CFRP), metal matrix composites (MMC), ceramics, graphite, titanium, printed circuit boards (PCB) and other hard-to-process materials [2]. The diamond-coated cutting tools have higher wear resistance and longer service life compared to uncoated ones. However, the characteristics of diamond films, both tribological and physical–mechanical, may have apparent differences in the perfection of the crystallite structure and the morphology of the surface of the diamond film. Nanocrystalline diamond (NCD) films, due to their record high hardness and reduced roughness, are particularly suitable for mechanical and tribological applications [5]. The primary purpose of diamond coatings applied to WC-Co tools is to ensure wear resistance and heat removal from the cutting edge; for this purpose, the films must be resistant to delamination and to chipping [6].

The technology of applying polycrystalline diamond coatings is quite complex, multi-stage, and has some technological features [7–10]. The most common method of applying DC to tungsten carbide-based hard alloys (WC-Co) is gas phase CVD deposition [11], more often using the hot filament chemical vapors deposition method (HF CVD) [12] or, less frequently, the microwave plasma chemical vapors deposition method (MP CVD); however, this provides a higher growth rate and film quality compared to HF CVD. The expansion of the possibilities of widespread use of MP CVD is hindered by a violation of the stability of the process and the uniformity of the diamond film, which depends on the shape and ratio of the 3D dimensions of the substrates and their aspect number [13,14], as well as the types of unique means to ensure uniform heating associated with the shape of the plasma [15]. The formation of the microwave plasma is sensitive to the geometry of the substrate due to the edge effect: the plasma is concentrated in areas of electric field distortion, especially at the edge of a strongly protruding substrate, which causes uneven deposition of the diamond film. The detrimental effect of the local microwave field on the temperature increase is aggravated with an increase in the substrate length [16,17], for example, for a drill or a mill. Therefore, temperature uniformity becomes a critical condition in determining the possibility of the successful growth of the diamond film.

In the work cycle [18–20], DC deposition on WC-Co was carried out at a substrate temperature of 700–750 °C and a pressure in the chamber of 80 Torr. The proportion of methane in hydrogen was 1%, the deposition time was six hours, and the deposition rate of DC was 1.5 μm/h. In Refs. [21–23], when nanocrystalline DC was deposited in microwave plasma, nitrogen was added to the gas mixture, and smooth and solid (80 GPa) DC ~10 μm thick was obtained. In Refs. [24–26], the samples of a new type of super hard tool with significantly higher wear resistance were created based on DC, intended for use in producing parts and structures made of difficult-to-process composite materials. To a large extent, the root problem of DC deposition in microwave plasma has been solved, due to the so-called edge effect, which consists of perturbation of the electric microwave field due to the heterogeneity of its concentration in the plasma volume and temperature along the perimeter of a metal substrate with a low aspect number when the substrate lies on the surface with the largest linear size on the central conducting platform. Therefore, the grain size inhomogeneities in the coating area, uneven coating thickness, and different degrees of inclusions of the non-diamond phase are possible, which is unacceptable for cutting tools used for high-precision surface treatment when an essential characteristic of which is the homogeneous fine-grained structure of DC.

In Refs. [24,25], an original design of the substrate holder was proposed, which enables to implementation of the deposition mode of a homogeneous DC and eliminates direct heating of low-aspect (height-to-diameter ratio) substrates in a microwave plasma reactor. The transition to indirect heating in the ethereal region of the plasma enabled the reduction in the edge effect sharply and protected the edges ether the substrates from overheating, as a result of which coatings with uniform morphology were obtained both in the center region and on the periphery of the DC. It was found that the temperature affecting the growth rate of diamond and the grain size of polycrystalline DC depends on the distance between the plasma-forming surface of the substrate holder and the growth surface of the substrate. The temperature range was determined with an accuracy of 20 °C. It was possible to deposit DC on plates made of WC-6% Co with an aspect ratio of 0.45. The plate's roughness in the center and periphery of the coating differed by no more than 10% [27]. It is shown that an effective method of regulating the temperature of the growth surface and controlling the structure of polycrystalline DC is an operational change in the distance between the substrate and the substrate holder during synthesis in microwave plasma with the position of the cooled substrate holder unchanged.

This paper presents a method for applying a conformal multilayer MCD/NCD coating in MPCVD plasma to an axial tool model (TM) with a high aspect number. The microwave source of the Ardis-100 reactor operates at a frequency of 2.45 GHz (wavelength $\lambda \approx 122$ mm). Therefore, an excessive ring (ER) with two types of holes ~1/2 and 3/4 λ was used to weaken the density of the microwave electric field acting directly on the TM volume inside the ring. The outer diameter of the ER is determined by taking into account the length of the TM to accommodate another part of the TM volume that acts as a holder and protects it from heating by microwave plasma. The undesirable effect of the edge is minimized. It is assumed that the primary heating source is indirect heating from the thermal radiation of high-temperature plasma generated in the region of the microwave field beam symmetrical to the inner hole of the ER. The TM heating during diamond deposition is modeled using a surface heater above the model with a lower surface temperature of 1400 °C. Modeling was performed using the finite element method (ANSYS) for TM with a diameter of 12 mm and a length of 75 mm inserted into a molybdenum sector of a 15 mm high ER with a hole for TM of 13.5 mm. It can be seen that this thermal system reaches stationary heat exchange conditions in about 20 s, and heating/cooling to T = 400 °C takes less than 5 min of rotation around its axis. Hence, the rotation speed is determined to be at least 12 rpm. The heating/cooling cycles in building up the diamond film should be shorter than 2.5 s with such a small thermal inertia of the model. The structure of the diamond film deposited with TM rotation is a 12-layer MCD/NCD conformal coating with a total length of 40 mm with an average NCD grain size of 51 nm.

2. Materials and Methods

The DC growth was carried out in the ARDIS-100 microwave reactor (Opticsystems, Moscow, Russia). The simulation was carried out in the ANSYS Multiphysics (Canonsburg, PA, USA) program (the steady-state thermal module for calculating thermal fields and heat transfer optimization options for leveling the tool model (TM) temperature, the transient thermal module for calculating the thermal inertia of the system and the minimum required rotation speed of the TM). The model is a cylindrical imitation of a tool with a diameter of 12 mm inserted into a molybdenum holder with a height of 15 mm with a hole for TM of 13.5 mm. The internal diameter of the molybdenum ER measured 52 and 75 mm. TM heating during diamond deposition is modeled using a surface heater with a surface temperature of 1400 °C.

A calibrated rod of KFM-39 tungsten carbide manufactured by Konrad Micro Drill (Kulmbach, Germany) hard alloy with a size of Ø12 × 75 mm was used as a model for coating, representing the equivalent of an axial cutting tool. A hole in one of the four sectors of the ER made of molybdenum was used as a TM holder. The region in the diametral and lateral sectors of the ER served for visual and temperature control of the TM heating. The rotation mode was carried out using a microwave transparent shaft from a stepper drive mechanism hermetically installed in the flange of the reactor.

The controlled TM surface was chemically etched with Murakami and Karo reagents, and to prevent cobalt diffusion at the deposition temperature of the coating, a 600 nm thick layer of tungsten was applied via magnetron sputtering based on the schematic diagram of obtaining diamond films on WC-Co alloy [7].

The TM surface was pre-coated with diamond nanoparticles about 5 nm in size by centrifugation (2000 rpm) from a water-based suspension. The growth of polycrystalline diamond films occurred in a methane/hydrogen gas mixture at a concentration of CH4/H2 = 4%, a pressure of 65 Torr, a microwave radiation power of 5 kW, and a substrate temperature of 800 °C ± 15 °C, which was controlled using a two-beam pyrometer METIS M322 (SensorTherm GmbH, Sulzbach, Germany). Additionally, 12-layer DC with reduced roughness was obtained according to the method described in [25]. The stimulation of secondary diamond nucleation was carried out via short-term nitrogen additions to the gas mixture during synthesis. The DC consists of alternating layers grown in microcrystalline diamond (H_2/CH_4 = 480/20 cm^3/min) modes for 25 min and nanocrystalline diamond ($N_2/CH_4/H_2$ = 20/20/460 cm^3/min) modes for 5 min. The total DC growth time was 6 h.

Microstructural studies and energy-dispersive X-ray spectroscopy were performed using a JEOL JSM-7001F microscope (Tokyo, Japan) at an electron energy of 30 keV to probe the material to a depth of approximately 3 μm.

The surface topology was investigated using the ZYGO NewView 5000 white light profilometer (Middlefield, CT, USA).

The uniformity of the coating on the surface of a TM was estimated through variations from the roundness of the profile and alignment by measuring deviations of the radius vector in the Cartesian coordinate system, followed by the calculation of approximating circles using the least squares method after applying the 50 irregularities/revolution filter. For that purpose, MarForm MMQ 400 (Göttingen, Germany) was used with a probe with a ball tip diameter of 3.0000 mm, and a contact force of 0.2 N. The profiler Hommel Tester T8000 (Hommelwerke Gmbh, Villingen-Schwenningen, Germany) was also used. The diameters of the workpiece were measured using the NORGAU NVMII-4030Di video measuring system (Moscow, Russia).

The structural phase analysis of the coating was performed via Raman spectroscopy using a LabRam HR 800 spectrometer (HORIBA Jobin Yvon, Tokyo, Japan) with the following parameters: laser radiation wavelength 473 nm, power 70 mW (the power on the sample was halved by the filter). The beam was focused using a long-focus lens Olympus ×100 (NA = 0.9) into a spot measuring Ø1 μm. A diffraction grating with a stroke density of 1800 mm^{-1} was used to measure the lines of the Raman spectra (RAMAN).

3. Results and Discussion

3.1. Simulation of a Cylindrical Tool Model Heating in Microwave Plasma (2.45 GHz)

Heat transfer between objects included radiation heating of surfaces from the lower surface of the heater with a temperature of 1400 °C, and gas convection in the internal volume of the model, as well as thermal conductivity in gas and solids. The cooling of the model was carried out by maintaining a constant temperature of the lower surface of the base at 25 C. Three calculations were carried out: one with the inner diameter of the molybdenum ring 1/2 λ and with the length of the TM part inside the molybdenum ring 40 mm, and two with the diameter of the ring 3/4 λ and the size of the protruding TM part 40 and 50 mm. The result of the thermal field calculation for the case of two (3/4 λ and 40 mm) is shown in Figure 1, from which it can be seen that for a good heat sink, the temperature of the molybdenum ring does not exceed 250 °C. The temperature of the TM varies significantly in different parts of the model depending on the efficiency of heat removal in the holder and in the free part of the TM. The surface temperature at the end and in the upper part of the TM is shown in the graphs in Figure 2.

From the graphs in Figure 2, it can be seen that an increase in the diameter of the molybdenum ER, as well as an increase in the length of the free end of the TM, contributes to the alignment of the thermal field both on the cylindrical surface and on the end of the TM. This phenomenon is due to the heat removal from the TM to the ring holder being 3–5 times higher than on the free surface of the cutter. This can be seen both through the large temperature gradient at the junction of the TM and the holder, and through the value of the heat flux on the surface of the TM. Additionally, in the conditions of non-stationary heat exchange, the thermal inertia of the model was evaluated with instantaneous switching on and off of the heater. The temperature graphs of the upper and lower parts of the cylindrical TM are shown in Figure 3.

It can be seen that this thermal system reaches stationary heat exchange conditions in about 20 s, and heating/cooling to T = 400 °C takes less than 5 s, that is, to smooth out the temperature difference on the TM surface by rotating around its axis. The rotation speed should be at least 12 rpm. With such a small the thermal inertia of the model, the heating/cooling cycles in building the diamond film should be shorter than 2.5 s.

The following design options of the ER and the placement of the TM were considered to reduce the longitudinal temperature gradient in the TM (Figure 4).

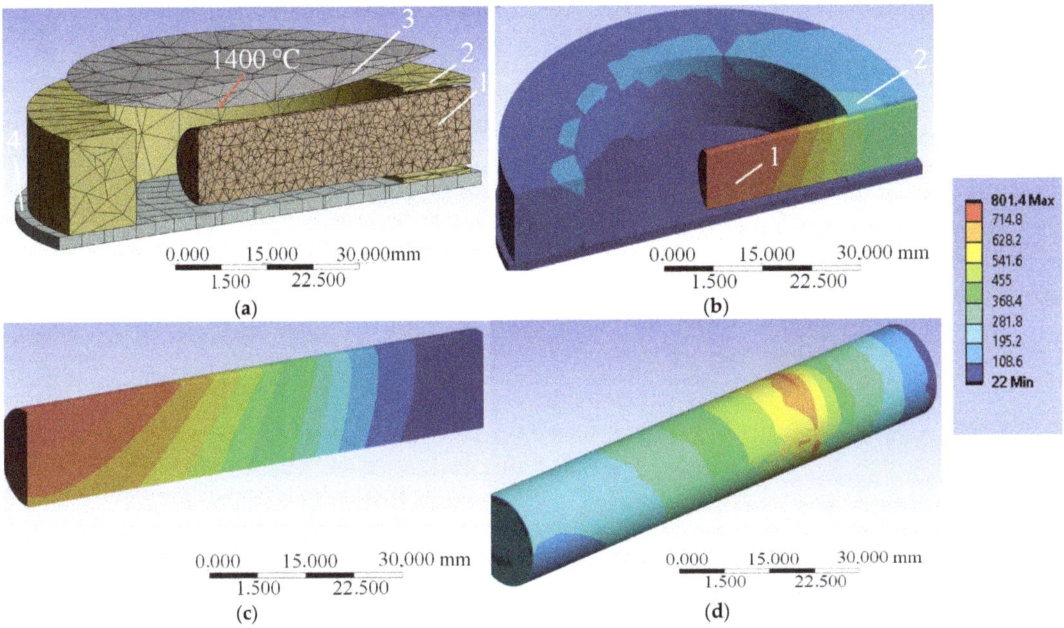

Figure 1. The results of modeling of the tool model (TM) heating: (**a**) the cross-section of the coating region broken down into finite elements: (1) the cylindrical axial model of the tool, (2) the sector of the excessive ring (ER) with the axis of the hole parallel to the central conductive platform, (3) the surface of the radiation heater with a temperature of 1400 °C, (4) the central conductive platform; (**b**) the cross-section of the thermal field in the modeling coating region (the heating isotherms of the tool model sections (1) and the excessive ring (2) are highlighted in color); (**c**) T = the thermal fields in the cylindrical TM without rotation; (**d**) the heat flow on the TM surface.

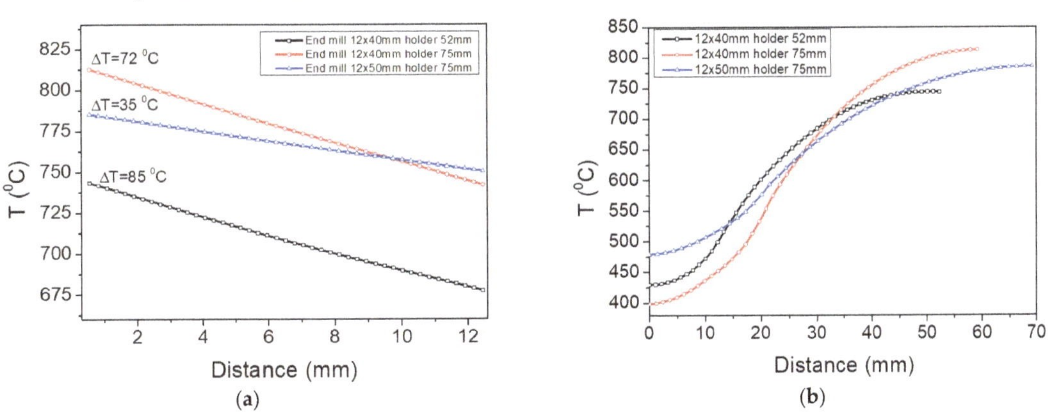

Figure 2. Temperature graphs at the end of the TM; (**a**) in the top-down direction, on the upper cylindrical surface of the model; (**b**) from the cold end to the hot end.

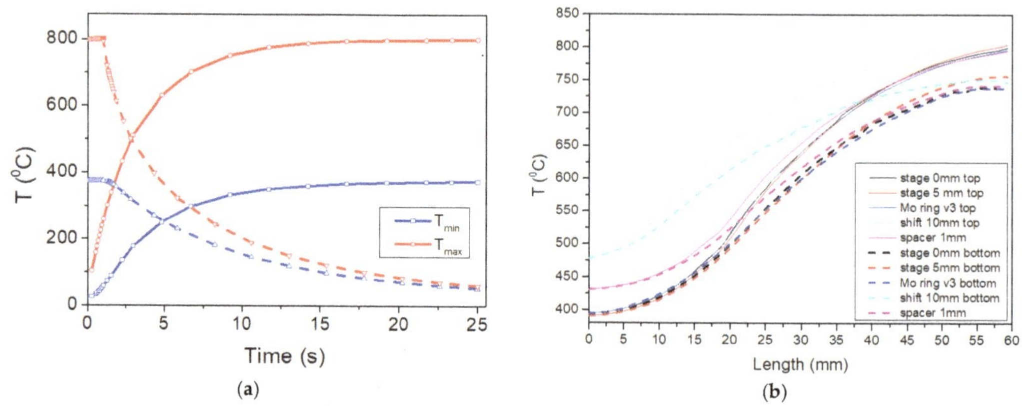

Figure 3. The temperature graphs of the upper and lower parts of the cylindrical TM, (**a**) heating curves (solid lines) and cooling curves (dotted lines) of the TM body with a sharp change in the heater operation mode; (**b**) the temperature of the lower and upper surfaces of the TM by size Ø12 × 60 mm for different versions of the working chamber geometry.

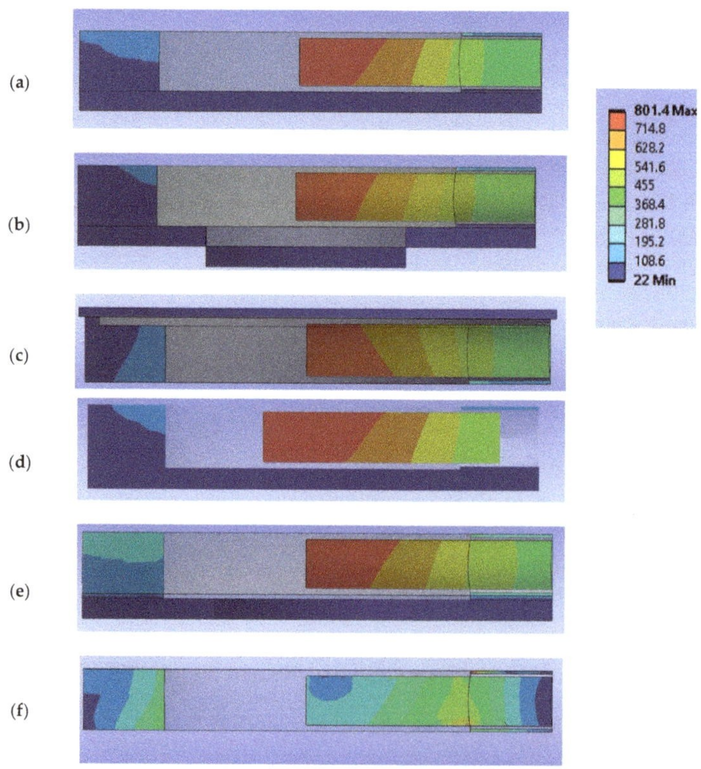

Figure 4. The results of modeling the tool model heating for the following cases: (**a**) the initial configuration; (**b**) lowering the movable base of the molybdenum excessive ring with a diameter of 50 mm by 5 mm down; (**c**) with a groove on the lower surface of the excessive ring so that the ring touches the base with an outer part 3.5 mm thick; (**d**) extending the tool model inside the excessive ring by 10 mm; (**e**) lining under the excessive ring of quartz gaskets with a thickness of 1 mm; (**f**) heat flow in the tool model body and the excessive ring.

The thermal fields were calculated for the considered configurations. It is concluded that it is necessary to minimize the heat sink through the molybdenum ER, in the version with a quartz gasket, and to reduce the thermal contact between the tool model body and the molybdenum ring since these places are the heat removal bridges in this thermal model. The best result is shown by moving the TM into the hole by 10 mm. Ideally, only the TM holder should remain inside the hole, and the heat sink through the holder should be excluded.

The results of the TM heating modeling showed that the temperature along the TM axis decreases from the free end to the end fixed in the holder inside the molybdenum ring. The TM body temperature drops sharply in the area near the holder. Rotation above 12 rpm during deposition makes it possible to compensate for the radial temperature unevenness on the TM surface. It is possible to balance the temperature difference along the axis only by increasing the length of the TM free part since heat dissipation in the holder is much higher, and changing its design does not lead to a fundamental change in the conditions of heat exchange. The departure inside the outer ring of the working part of the TM with a temperature of 700–800 °C, in which the growth of the diamond coating is possible, is 20–25 mm.

For experimental verification of the simulation results, the following scheme of a growth plant with an Ardis-100 microwave reactor was adopted (Figure 5).

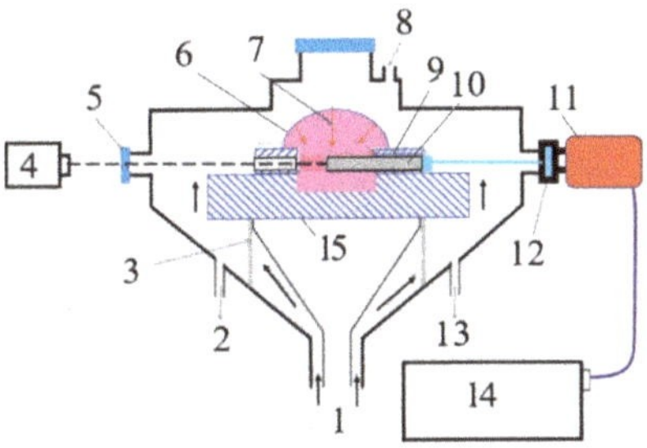

Figure 5. The scheme of a growth plant with an Ardis-100 microwave reactor: 1—microwave radiation; 2—pumping; 3—a window for microwave; 4—pyrometer; 5—viewing window; 6—plasma mode E_{01} area; 7—radiation heating; 8—gas supply; 9—molybdenum holder (exorbitant ring); 10—tool model (TM); 11—tool model rotation mechanism; 12—flange; 13—pumping; 14—tool model rotation control; 15—conducting platform.

The results of modeling the temperature distribution over TM agree with the experimental data related to pyrometric temperature measurement on the workpiece (Figure 6).

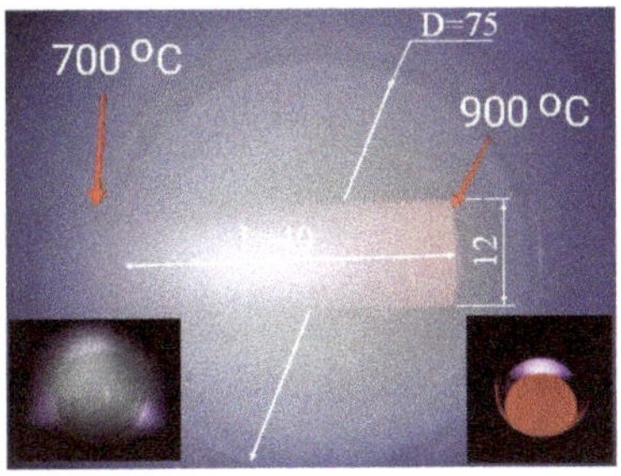

Figure 6. A photograph of the growth of a multilayer MCD/NCD diamond film with tool model rotation (Ø12 mm, D = 3/4 λ, where λ is the wavelength for microwave 2.45 GHz), viewed through the upper window; in the insert on the left, a photo of the growth of a diamond film without TM rotation, viewed through the upper window, D = 1/2λ; in the insert on the right, there is a photo of the end of the TM during the growth of the diamond film, viewed through the side window.

3.2. Results of Linear Measurements of a Cylindrical Workpiece Coated with Polycrystalline Diamond

The measurement results are shown in Figure 7 and Table 1. The measurement positions of the diameters and deviations from the circles coincide. The eccentricity value was 2.09 μm.

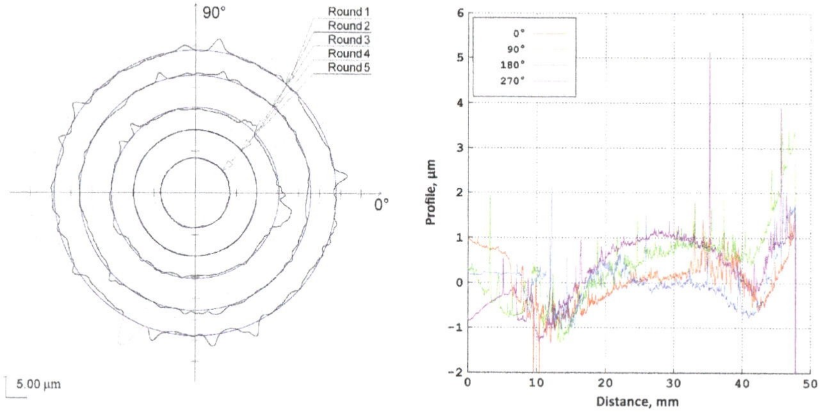

(a) (b)

Figure 7. Circular frames (**a**) end profiles along the lines along the forming cylinder (**b**) of TM with diamond coated with rotation.

Table 1. Values of linear measurements of cylindrical TM coated with polycrystalline diamond.

Round	Distance from the End of the Workpiece, mm	Diameter, mm	Deviation from Roundness, μm	Note
1	1	11.9967	4.34	covered area
2	10	11.9961	2.86	covered area
3	20	11.9919	4.09	covered area
4	40	11.9918	0.44	the area with etching without coating
5	50	11.9940	0.56	initial workpiece

It can be seen from Figure 7 that the single high values of deviations from the circumference and the profiles along the lines along the forming cylinder, as well as the presence of eccentricity, are caused by the chaotic distribution over the surface of individual elements of the growth of the diamond coating up to 5 μm high. Without them, the deviation does not exceed 1 μm. The coating thickness is uniform. According to Table 1, the variation from the roundness of the surface of the uncoated TM at a distance of 50 mm is 0.56 μm. In the coated areas, two maxima are distinguished at a distance of 1 and 20 mm from the end of 4.34 and 4.09 μm, respectively, and between them, a decrease to 2.86 μm.

The average surface roughness values Ra $\approx$ 5 μm and Rz up to 27 μm were measured with an optical profilometer on an area of 0.9 $\times$ 0.7 mm for the initial samples. The surface reliefs of the films for the lines along the forming cylinders of the TM in four positions through 90° with a step of 10 mm are compared in Figure 8. At the end and at a distance of 10 mm from the rear, areas with increased roughness values were observed.

The analysis shows that the regularity of the increase in roughness along one of the measured lines at any distance from the end is not observed. The most significant values are marked on the 180° degree line at a distance of 1 mm from the rear and on the 0°, and 90° lines at a distance of 10 mm; the size of the image area was chosen in so as to obtain sufficient statistics.

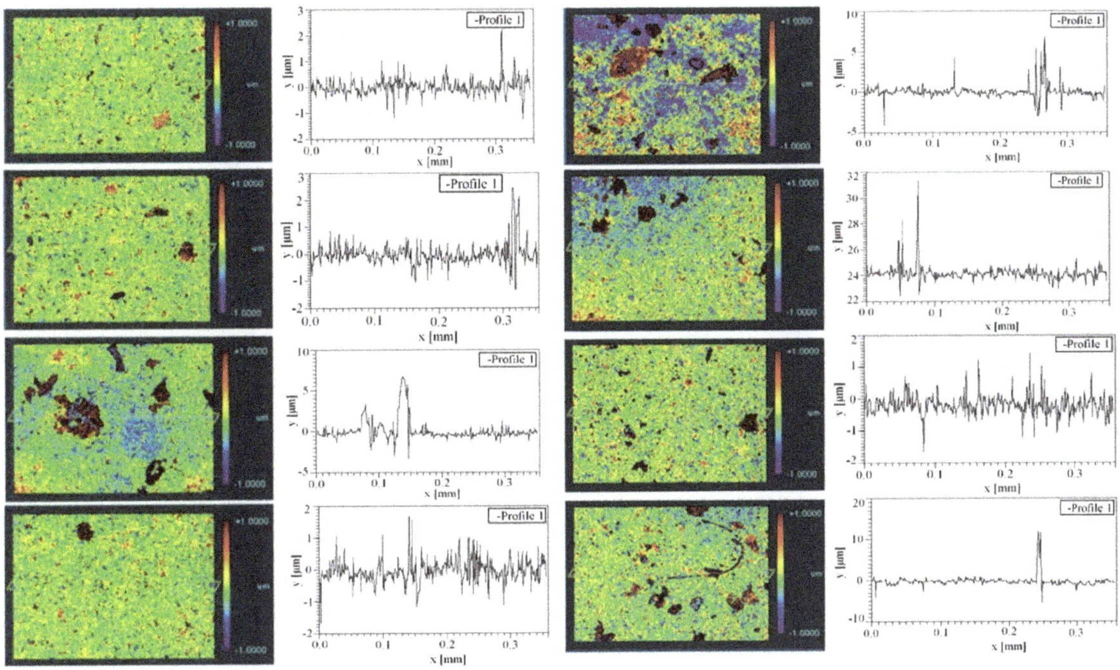

(**a**) (**b**)

Figure 8. The relief of the TM surface deposited within 5 h, measured using an optical profilometer on an area of 0.36 mm $\times$ 0.27 mm: (**a**) the relief of the surface formed at a distance of 1 mm from the TM end along the perimeter of rotation; (**b**) at a distance of 10 mm from the TM end. The left panels are 2D profiles with pseudo-height color coding. The right panels are linear 2D profiles along randomly selected straight lines.

A more pronounced smoothing from Ra $\approx$ 186, 222, and up to Ra $\approx$ 287 nm was detected after measuring the end face on the line 0°,90°, and 270° (Figure 8a), and from Ra $\approx$ 222, 238 and 1270 nm at a distance of 10 mm on the line 90°, 180°, and 270° (Figure 8b). Since the deposition of diamond material occurs via rotation at the calculated speed, the final roughness cannot be equalized in a narrower interval. Improved surface roughness can be achieved using alternative rates found experimentally in further studies.

3.3. Results of RAMAN Spectra

The spectra for Raman analysis were recorded along the lines starting from the end every 10 mm. The lines are located on the cylinder on opposite sides to each other. That is, each line is 90° away from the previous one. Thus, we obtained four lines of four spectra in each. The spectra (Figure 9) are typical for a nanocrystalline CVD diamond film with a highly disordered graphite sp^2 phase content and CH-compounds. The presence of graphite is indicated by D and G bands located at 1350 and 1550 cm^{-1}, respectively. These bands have a significantly higher intensity than the diamond peak about 1337 cm^{-1}. The high CH compounds' content at the intercrystalline boundaries is indicated by trans polyacetylene (TPA) bands at 1140 and 1490 cm^{-1} due to the uneven heating of the milling cutter. Significant differences in the shape of the spectra were observed at points recorded at the same distance from the end face but on different lines due to the uneven heating of the milling cutter.

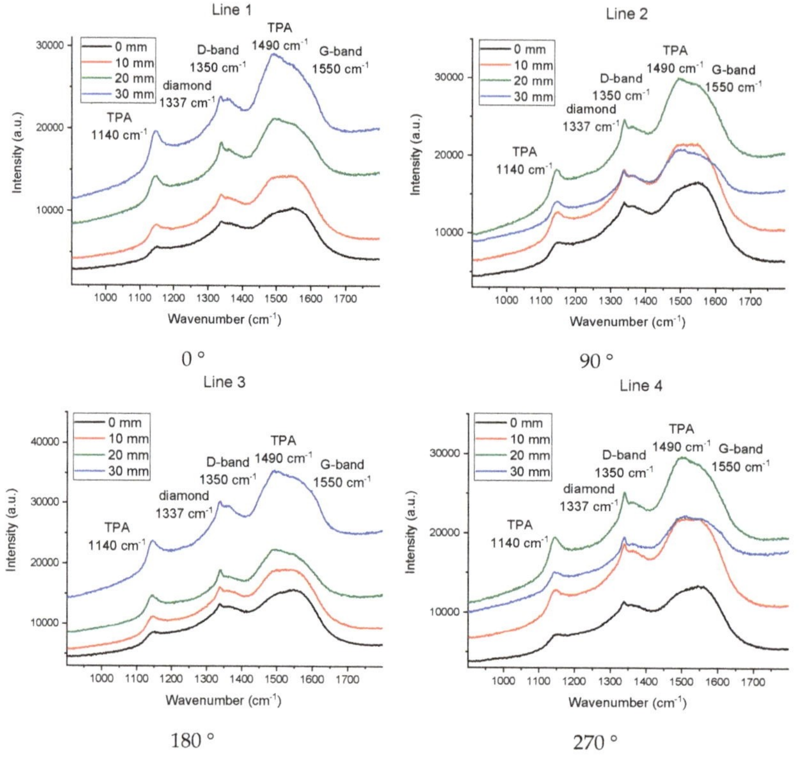

Figure 9. Raman spectra along four lines of the forming cylinder through 90 °, at a distance of 1, 10, 20, and 30 mm from the end of the TM.

The parameters of these bands such as intensity, position, and width at half-height were determined by approximating the spectra with Lorentz curves (for the diamond peak and the trans polyacetylene band) and Gauss curves (for the G band). Figure 10a shows graphs of the dependence of the width of the diamond peak on the coordinate. The peak width is directly related to the size of diamond crystallites. It was more than 12 cm^{-1}, which is typical for nanocrystalline diamonds, at all points of the milling cutter. The maximum width of the diamond peak (~15.5 cm^{-1}) is reached at a distance of 10 mm from the end and the minimum (~13.0 cm^{-1})—at a length of 30 mm. A similar pattern is observed for the width of the TPA strip at 11.4 cm^{-1} (Figure 10b), which is strictly reduced from ~50 cm^{-1} at the end of the cutter to ~40 cm^{-1} at a distance of 30 mm from the rear. Thus, the lower the substrate temperature, the larger, on average, the crystallites of the nanocrystalline diamond film.

The diamond film experiences elastic compression stresses at all points judging by the position of the diamond peak Figure 10c. During the growth of the film, the milling cutter was heated, and when cooled after synthesis, it decreased in size; thus, the diamond coating was compressed. Judging by the dependence graph, the value of elastic stresses reaches a maximum at the end of the milling cutter since this area was the most heated during the synthesis process. Figure 10d shows the ratio between the diamond peak areas and the G-band of disordered carbon. There is a tendency to decrease the proportion of the disordered phase compared to the diamond when moving away from the end of the cutter. It is especially clearly manifested in line 4. At a distance from the end, the temperature of the substrate was lower. As a result, the film growth was slower but with a higher percentage of diamond.

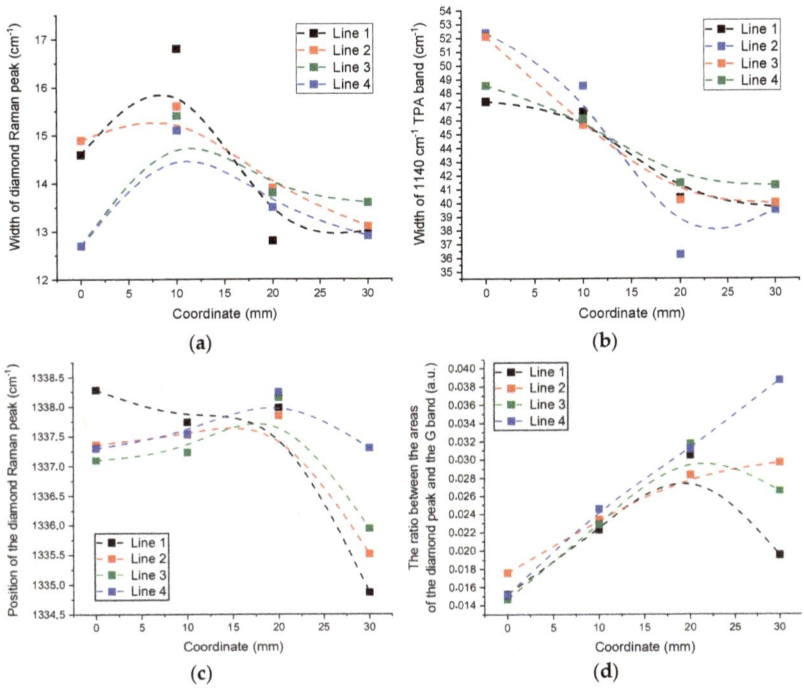

Figure 10. Characterization of the diamond film depending on the coordinate: (**a**) The width of the peak of the diamond CR depending on the coordinate; (**b**) the width of the trans polyacetylene strip by 1140 cm^{-1} depending on the coordinate; (**c**) the position of the peak of the diamond CR depending on the coordinate. (**d**) the ratio between the areas of the diamond peak and the G-band depends on the coordinate.

3.4. Morphology of the Diamond Film Surface

The morphology of the diamond film surface on a cylindrical MV sample with rotation is shown in Figure 11. The average diamond grains sizes are presented in Table 2. It was determined from SEM photographs using the Thixomet Pro (St Petersburg, Russia) automatic image analysis system for the microstructure of metals and alloys via quantitative metallography.

According to Table 2, a graph of the dependence of the average grain size on DC film on the lines 0°–90°–180°–270° along the forming TM with a step of 10 mm was constructed (Figure 12).

In all cases, the maximum grain sizes are observed on the cylinder sample with rotation (Table 2) at a distance of 30 mm from the end (the average grain size at points 1–4, 2–4, 3–4, 4–4 d = 73 ± 4 nm); the minimum are observed at the end of the cutting edge (d = 51 ± 7 nm at points 1–1, 2–1, 3–1, 4–1); and between them we observed intermediate values: d = 58± 7 nm at a distance of 10 mm from the end at points 1–2, 2–2, 3–2, 4–2, and d = 62 ± 7 nm at a distance of 20 mm from the end at points 1–3, 2–3, 3–3, 4–3. The average grain size at a distance of 10 mm from the end is 13% larger than at the end; similarly, for 20 mm, it is 20%, and for 30 mm, it is 41%. The average grain sizes grow slower as they move away from the end face, which correlates with the local temperature of the sample surface or, in another way, with the distance from the center of the plasma cloud during diamond growth, where the maximum temperature is observed. In the case of a cylinder sample without rotation at the end of the cutting edge, both the average grain size and the size spread are smaller: d = 45 ± 3 nm (at similar points 1–1, 2–1, 3–1, 4–1), while there is no diamond in volume 3–3 on the cylinder without rotation. The phase is a nanocrystalline graphite film; in SEM, it is indistinguishable from a nanocrystalline diamond film.

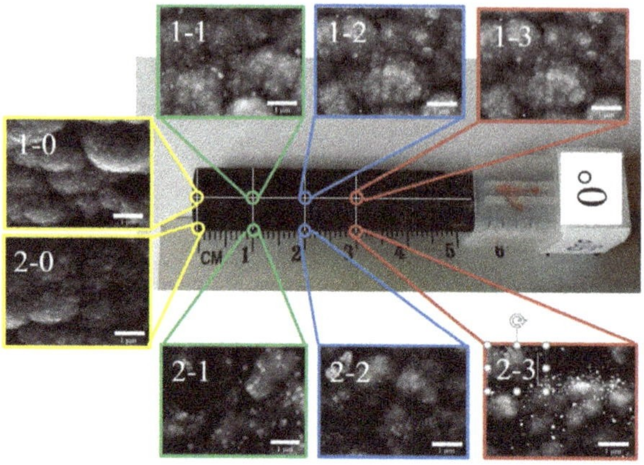

(a)

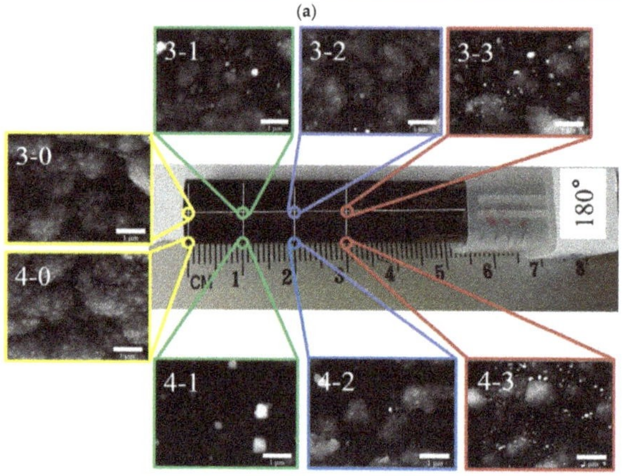

(b)

Figure 11. Morphology of the diamond film surface along lines starting from the end every 10 mm on a cylindrical sample deposited in MP CVD with rotation: (**a**) SEM images of measurement points along two lines of the cylinder forming at angles of 0°–90°; (**b**) SEM images of measurement points along two lines of the cylinder forming at angles 180° and 270° designations (Xi-Yi) Xi-angle; Yi-distance from the edge along the line forming the cylinder, scale mark 1 µm.

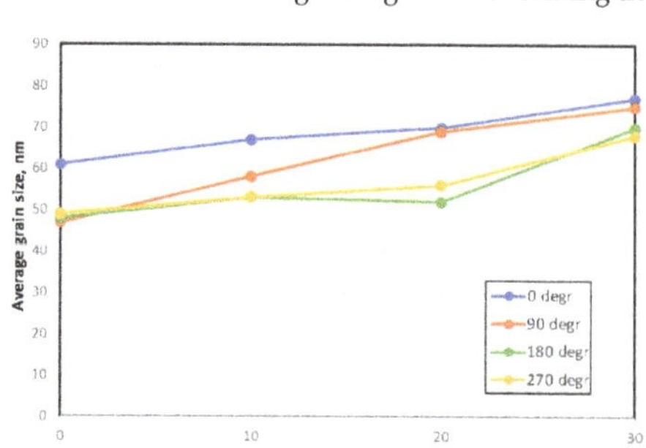

Figure 12. The dependence of the average grain size on the CD film on the lines 0°–90°–180°–270° along the forming tool model in increments of 10 mm.

Table 2. The average sizes of diamond grains.

	The Measured Point on the Surface of the Diamond Film	Average Grain Size nm, on a Cylindrical Sample without Rotation	Average Grain Size nm, on a Cylindrical Sample with Rotation
1–0	end on top	49 ± 9	61 ± 16
1–1	10 mm from the top end		67 ± 15
1–2	20 mm from the top end		70 ± 15
1–3	30 mm from the top end		77 ± 15
2–0	end on the right	41 ± 4	47 ± 12
2–1	10 mm from the end on the right		58 ± 21
2–2	20 mm from the end on the right		69 ± 15
2–3	20 mm from the end on the right		75 ± 12
3–0	bottom end	44 ± 6	48 ± 15
3–1	10 mm from the bottom end		53 ± 10
3–2	20 mm from the bottom end		52 ± 10
3–3	30 mm from the bottom end		70 ± 20
4–0	end on the left	46 ± 9	49 ± 10
4–1	10 mm from the end on the left		53 ± 13
4–2	10 mm from the end on the left		56 ± 14
4–3	10 mm from the end on the left		68 ± 17

Figure 13 shows the morphology of a diamond film formed at the end of a cylindrical tool model under conditions without rotation. The dependence of the average grain size on the angle of rotation when measuring TM at the end without rotation is presented in Table 3.

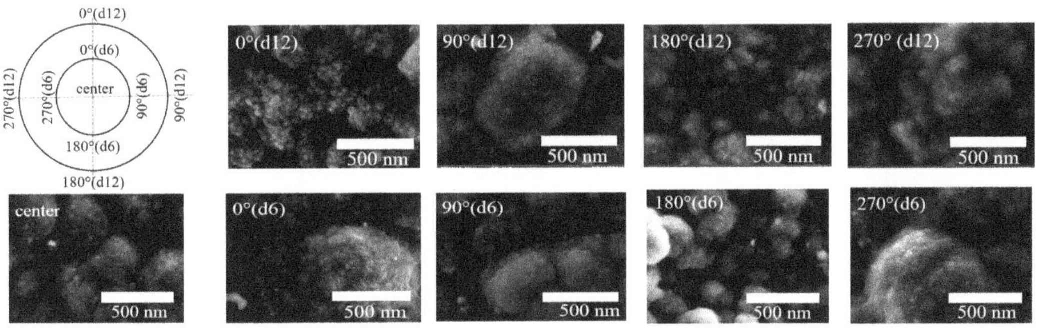

Figure 13. Morphology of the surface of the NCD film at the end of a carbide cylindrical sample deposited in MP CVD without rotation: SEM images of measurement points on the diameters d12, d6, and in the center, rotation angle when measured via 90° clockwise rotation, scale mark 500 nm.

Table 3. The average grain size at the end of TM without rotation.

Distance from the Center of the End Face (mm)	The Angle of Rotation, °	Average Grain Size Without Rotation (nm)
0	-	37 ± 4
6	0	44 ± 6
	90	32 ± 7
	180	41 ± 4
	270	34 ± 8
12	0	49 ± 9
	90	41± 4
	180	44 ± 6
	270	46 ± 9

According to Table 3, a graph of the average grain size of the TM at the no-rotation mode is shown in Figure 14. Here, the dimensions of the diamond grains at distances of 6 and 12 mm from the center of the TM end depending on the angles of location (0°, 90°, 180°, 270°) relevant to the points indicated are presented, according to Table 3. The maximum value of the diamond grain size is 49 nm, and the face is located 12 mm from the center. The minimum value of the diamond grain is 32 nm and it is at a distance of 6 mm from the center of the end face. In the center of the rake face, the size of the diamond grain is 37 nm.

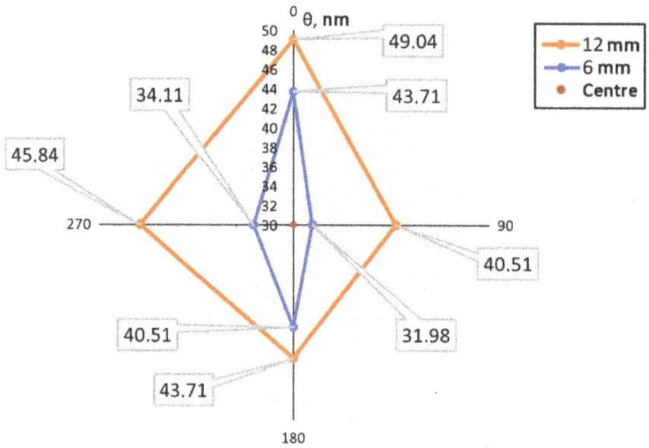

Figure 14. The average grain size on the tool model along the end face is without rotation mode.

4. Conclusions

Mathematical modeling has determined that the temperature along the cylindrical tool model (TM) axis decreases from the free end to the one fixed in the holder inside the molybdenum ring. The TM body cools down especially sharply in the area near the holder.

It is possible to reduce the temperature difference along the axis by increasing the length of the free part of the TM since heat dissipation in the holder is much higher, and changing its design does not lead to a fundamental change in the heat exchange conditions. The departure of the TM part with a temperature of 700–800 °C, in which the growth of the diamond coating without rotation is possible in the ARDIS-100 type reactor, is 20–25 mm.

Rotation above 12 rpm during deposition makes it possible to compensate for the radial temperature unevenness on the TM surface and increase the growth of the diamond coating to 40 mm.

In all cases, in the carbide cylinder sample with rotation, the maximum grain sizes are observed at a distance of 30 mm from the end (the average grain size is 73 ± 4 nm); the minimum are observed at the back at the cutting edge, 51 ± 7 nm; between them, intermediate values are observed: 58 ± 7 nm at a distance of 10 mm from the rear, and 62 ± 7 nm at a distance of 20 mm from the rear.

The average grain size at a distance of 10 mm from the end is 13% larger than at the rear; similarly, at a distance of 20 mm, it is 20%, and a distance of 30 mm, it is 41%.

The average grain sizes grow slower as they move away from the end face, which correlates with the local temperature of the sample surface or with the distance from the center of the plasma cloud, where the maximum temperature is observed during diamond growth.

In the case of a sample of the carbide cylinder without rotation at the end, both the average grain size and the size spread are smaller—45 ± 3 nm. At the same time, there is no diamond phase on the underside of the cylinder. The phase is a nanocrystalline graphite film; however, in SEM, it is indistinguishable from a nanocrystalline diamond film.

The observed variation in the values of roughness and grain size can be attributed to the TM rotation speed 12 min^{-1}, which was determined via mathematical modeling and required experimental adjustment.

Author Contributions: Conceptualization, V.I.K. and E.E.A.; methodology, D.N.S. and A.K.M.; investigation, E.V.Z. and A.P.L.; validation, V.S.S. and A.A.K.; steady-state thermal and transient thermal modelling, A.F.P.; formal analysis, V.Y.Y., A.P.B. and S.G.R.; project administration, S.N.G.; writing and editing S.V.F. All authors have read and agreed to the published version of the manuscript.

Funding: This research was carried out at the expense of the grant of the Russian Science Foundation No. 22-19-00694.

Institutional Review Board Statement: Not applicable.

Informed Consent Statement: Not applicable.

Data Availability Statement: The data presented in this study are available on request from the corresponding author.

Acknowledgments: This study was carried with on the equipment of the Center of collective use of MSUT "STANKIN".

Conflicts of Interest: The authors declare no conflict of interest.

References

1. Gielisse, P.J. Mechanical properties of diamond, diamond films, diamond-like carbon and diamond materials. In *Handbook of Industrial Diamonds and Diamond Films*; Prelas, M.A., Popovici, G., Bigelow, L.K., Eds.; CRC Press: New York, NY, USA, 1998; pp. 49–89.
2. Tillmann, W. Trends and market perspectives for diamond tools in the construction industry. *Int. J. Refract. Met. Hard Mater.* **2000**, *18*, 301–306. [CrossRef]
3. Polini, R. Chemically vapour deposited diamond coatings on cemented tungsten carbides: Substrate pretreatments, adhesion and cutting performance. *Thin Solid Films* **2006**, *515*, 4–13. [CrossRef]
4. Li, Y.S.; Tang, Y.; Yang, Q.; Shimada, S.; Wei, R.; Lee, K.Y.; Hirose, A. Al-enhanced nucleation and adhesion of diamond films on WC-Co substrates. *Int. J. Refract. Met. Hard Mater.* **2008**, *26*, 465–471. [CrossRef]
5. Wang, C.; Wang, X.; Sun, F. Tribological behavior and cutting performance of monolayer, bilayer and multilayer diamond coated milling tools in machining of zirconia ceramics. *Surf. Coat. Technol.* **2018**, *353*, 49–57. [CrossRef]
6. Silva, E.L.; Pratas, S.; Neto, M.A.; Fernandes, C.M.; Figueiredo, D.; Silva, R.F. Multilayer Diamond Coatings Applied to Micro-End-Milling of Cemented Carbide. *Materials* **2021**, *14*, 3333. [CrossRef]
7. Ashkinazi, E.; Fedorov, S.; Khomich, A.; Rogalin, V.; Bolshakov, A.; Sovyk, D.; Grigoriev, S.; Konov, V. Technology Features of Diamond Coating Deposition on a Carbide Tool. *C* **2022**, *8*, 77. [CrossRef]
8. Sedov, V.; Martyanov, A.; Ashkinazi, E.; Tiazhelov, I.; Savin, S.; Sovyk, D.; Mandal, S.; Fedorov, S.; Grigoriev, S.; Ralchenko, V. Effect of diamond seeds size on the adhesion of CVD diamond coatings on WC-Co instrument. *Surf. Interfaces* **2023**, *38*, 102861. [CrossRef]
9. Haubner, R.; Kalss, W. Diamond deposition on hard metal substrates—Comparison of substrate pre-treatments and industrial applications. *Int. J. Refract. Met. Hard Mater.* **2010**, *28*, 475–483. [CrossRef]
10. Fronzi, M.; Bishop, J.; Assadi, M.H.N.; Martin, A.A.; Assadi, M.; Regan, B.; Stampfl, C.; Aharonovich, I.; Ford, M.J.; Toth, M. Role of knock-on in electron beam induced etching of diamond. *Carbon* **2020**, *164*, 51–58. [CrossRef]
11. Kolber, T.; Koepf, A.; Haubner, R.; Hutter, H. WC-Co substrate surface pretreatments with aluminum compounds prior to polycrystalline CVD diamond deposition. *Int. J. Refract. Met. Hard Mater.* **1999**, *17*, 445–452. [CrossRef]
12. Dumpala, R.; Chandran, M.; Ramachandra Rao, M.S. Engineered CVD Diamond Coatings for Machining and Tribological Applications. *J. Miner. Met. Mater. Soc.* **2015**, *67*, 1565–1577. [CrossRef]
13. Wang, L.L.; Shen, B.; Sun, F.; Zhang, Z. Effect of boron-doped diamond interlayer on cutting performance of diamond coated micro drills for graphite machining. *Materials* **2013**, *6*, 3128–3138.
14. Rifai, A.; Creedon, D.; Tran, N.; Hejazi, M.; Garrett, D.; Greentree, A.D.; Pirogova, E.; Stacey, A.; Fox, K. Highly uniform polycrystalline diamond coatings of three-dimensional structures. *Surf. Coat. Technol.* **2021**, *408*, 126815. [CrossRef]
15. Nad, S.; Gu, Y.; Asmussen, J. Growth strategies for large and high-quality single crystal diamond substrates. *Diam. Relat. Mater.* **2015**, *60*, 26–34. [CrossRef]
16. Yamada, H.; Chayahara, A.; Mokuno, Y.; Soda, Y.; Horino, Y.; Shikata, S. Simulation of microwave plasmas concentrated on the top surface of a diamond substrate with finite thickness. *Diam. Relat. Mater.* **2006**, *15*, 1383–1388. [CrossRef]
17. Fernandes, A.J.S.; Silva, V.A.; Carrapichano, J.M.; Dias, G.R.; Silva, R.F.; Costa, F.M. MPCVD diamond tool cutting-edge coverage: Dependence on the side wedge angle. *Diam. Relat. Mater.* **2001**, *10*, 803–808. [CrossRef]
18. Xu, Z.; Lev, L.; Lukitsch, M.; Kumar, A. Effects of surface pretreatments on the deposition of adherent diamond coatings on cemented tungsten carbide substrates. *Diam. Relat. Mater.* **2007**, *16*, 461–466. [CrossRef]
19. Xu, Z.; Kumar, A.; Lev, L.; Lukitsch, M.; Sikder, A. Adhesion improvement of CVD diamond coatings on WC-Co substrates for machining applications. *Mater. Res. Soc. Symp.* **2005**, *843*, T3.33. [CrossRef]

20. Xu, Z.; Lev, L.; Lukitsch, M.; Kumar, A. Deposition of adherent diamond coatings on WC-Co substrates. *Mater. Res. Soc. Symp. Proc.* **2006**, *890*, 0890-Y01-02. [CrossRef]
21. Hu, J.; Chou, Y.K.; Thompson, R.G.; Burgess, J.; Street, S. Characterizations of nano-crystalline diamond coating cutting tools. *Surf. Coat. Technol.* **2007**, *202*, 1113–1117. [CrossRef]
22. Hu, J.; Chou, Y.K.; Thompson, R.G. Nanocrystalline diamond coating tools for machining high-strength Al alloys. *Int. J. Refract. Met. Hard Mater.* **2008**, *26*, 135–144. [CrossRef]
23. Chou, Y.K.; Thompson, R.G.; Kumar, A. CVD-diamond technologies for dry drilling applications. *Thin Solid Films* **2010**, *518*, 7487–7491. [CrossRef]
24. Ashkinazi, E.E.; Sedov, V.S.; Petrzhik, M.I.; Sovyk, D.N.; Khomich, A.A.; Ralchenko, V.G.; Vinogradov, D.V.; Tsygankov, P.A.; Ushakova, I.N.; Khomich, A.V. Effect of crystal structure on the tribological properties of diamond coatings on hard-alloy cutting tools. *J. Frict. Wear* **2017**, *38*, 252–258. [CrossRef]
25. Ashkihazi, E.E.; Sedov, V.S.; Sovyk, D.N.; Khomich, A.A.; Bolshakov, A.P.; Ryzhkov, S.G.; Khomich, A.V.; Vinogradov, D.V.; Ralchenko, V.G.; Konov, V.I. Plateholder design for deposition of uniform diamond coatings on WC-Co substrates by microwave plasma CVD for efficient turning application. *Diam. Relat. Mater.* **2017**, *75*, 169–175. [CrossRef]
26. Ashkinazi, E.E.; Ralchenko, V.G.; Konov, V.I.; Vinogradov, D.V.; Tsygankov, P.A.; Dryzhak, E.A.; Khomich, A.V. Frictional coefficients between aluminumsilicon alloy and cutting inserts with MPCVD diamond coatings. *Russ. Eng. Res.* **2018**, *38*, 457–461. [CrossRef]
27. Sedov, V.; Martyanov, A.; Savin, S.; Zavedeev, E.; Kudryavtsev, O.; Bland, H.; Mandal, S.; Williams, O.; Ralchenko, V.; Konov, V. CVD synthesis of multi-layered polycrystalline diamond films with reduced roughness using time-limited injections of N2 gas. *Diam. Relat. Mater.* **2021**, *114*, 108333. [CrossRef]

coatings MDPI

Article

Influence of Silver-Coated Tool Electrode on Electrochemical Micromachining of Incoloy 825

Geethapriyan Thangamani [1], Muthuramalingam Thangaraj [2,*], Palani Iyamperumal Anand [1], Mani Jayakumar [3], Nikolaos E. Karkalos [4,*], Emmanouil L. Papazoglou [4] and Panagiotis Karmiris-Obratański [5,*]

1. Mechatronics and Instrumentation Laboratory, Department of Mechanical Engineering, Indian Institute of Technology Indore, Khandwa Road, Simrol, Indore 453552, India
2. Department of Mechatronics Engineering, College of Engineering and Technology, SRM Institute of Science and Technology, Kattankulathur 603203, India
3. Electroplating and Metal Finishing Division, CSIR Central Electrochemical Research Institute (CSIR CECRI), Karaikudi 630006, India
4. Laboratory of Manufacturing Technology, School of Mechanical Engineering, National Technical University of Athens, 15780 Athens, Greece
5. Department of Manufacturing Systems, Faculty of Mechanical Engineering and Robotics, AGH University of Science and Technology, 30-059 Cracow, Poland

* Correspondence: muthurat@srmist.edu.in (M.T.); nkark@mail.ntua.gr (N.E.K.); karmiris@agh.edu.pl (P.K.-O.)

Abstract: Incoloy 825 alloy is often used in calorifiers, propeller shafts, and tank vehicles owing to the improved resistance to aqueous corrosion. The electrochemical micromachining process can be utilized to machine such an engineering material owing to higher precision and lower tool wear. In the present study, an investigation was performed to enhance the process of creating micro-holes using silver-coated copper tool electrodes. The sodium nitrate electrolyte was used under different levels of input parameters such as voltage, electrolyte concentration, frequency, and duty cycle with a view to improving material removal rate, conicity, overcut, and circularity. It was found that silver-coated copper tool electrode had a high material removal rate (MRR), better overcut, conicity, and circularity compared to uncoated copper tools in most cases, due to its high corrosive resistance and electrical conductivity. From SEM and EDS analysis, it was observed that better surface topography of the micro-holes is obtained with silver-coated copper tool electrode while machining Incoloy 825 alloy in the micromachining process.

Keywords: incoloy 825; ECMM; silver coating; overcut; MRR; conicity; circularity

1. Introduction

Incoloy 825 alloy is a nickel-iron-chromium alloy with additions of titanium, copper, and molybdenum often used in heat exchangers, propeller shafts, and tank vehicles owing to the improved resistance to aqueous corrosion [1]. Superalloys based on nickel, in contrast to those based on iron, do not require carbon to undergo microstructural changes in order to achieve their mechanical properties [2]. For superalloys to be able to achieve exceptional mechanical properties under service conditions, they must contain large amounts of alloying elements in their composition [3]. Incoloy 825, the austenitic matrix is embedded with Ti-rich precipitates that have a typical face-centred cubic microstructure (FCC) [4]. As a result of the precipitates of ordered Ni_3Al in a solid solution-strengthened nickel matrix, nickel-based superalloys, such as Incoloy 825 exhibit excellent high-temperature strength [5,6].

Given these exceptional material properties, superalloys are hard-to-machine with conventional methods. Thus, it is essential to introduce unconventional machining processes for these difficult-to-machine materials. Electrochemical Micromachining(ECMM) can be used to produce parts with specified shapes and dimensions accurately since it is best suited for machining micro-level profiles with better precision and low wear. In order to achieve these goals, it is important to enhance the process by appropriately choosing ECMM input parameters with respect to the workpiece material as well as introducing suitable modifications on selected tools and electrolytes.

In the relevant literature, a considerable amount of work has been conducted on the determination of appropriate ECMM parameter values. Electrochemical micromachining involves many responses and process parameters, such as supply voltage(VS), the concentration of electrolyte(CE), duty cycle(DC), and feed rate(FR); thus it is required that the choice of appropriate values for these parameters should be carried out appropriately. It was shown that ECM is one of the few most advanced micromachining processes that can handle a wide variety of applications, which can provide better MRR with low overcut [7]. ECMM process can make advanced and complex shapes in aeronautical industries and its capability to machine macro and micro features for biomedical applications was also found in some studies [8–10]. The current efficiency during ECMM was correlated with the current density, concentration, temperature, and flow velocity of the electrolyte [11].

Li et al. [12] performed through-mask electrochemical machining in order to produce micro-holes on Ti-6Al-4V sheets by using 10% $NaNO_3$ electrolyte and various combinations of voltage, pulse frequency, and duty factor. They found that the accuracy of holes was considerably improved at the optimum settings. Harugade, Kavade, and Hargude [13] investigated the effect of electrolytes on MRR during the ECMM process on a glass workpiece. Two types of electrolyte solution, namely KOH and H2SO4 were used under different values of voltage, electrolyte concentration, and inter-electrode gap. It was found that applied voltage was the most important parameter, followed by electrolyte concentration and KOH electrolyte exhibited the highest MRR. Taguchi Grey relational analysis was used to optimize the machining parameters such as feed rate, the discharge rate of electrolyte, and electrolyte concentration during electrochemical micro-machining of Inconel 625 with NaCl as electrolyte [14]. The performance parameters were MRR, Ra, overcut, circularity error, and perpendicularity error. The increase of all parameters leads to higher MRR, with feed rate being the most important parameter. Moreover, regarding Ra, the increase of feed rate and electrolyte concentration, as well as the decrease of flow rate contribute to lower Ra, with discharge rate being the most important parameter, whereas overcut was reduced at higher feed rates and lower flow rate and electrolyte concentrations, with feed rate being the most important parameter. Finally, a higher feed rate improved both circularity and perpendicularity errors.

Krishnan et al. [15] conducted a thorough study regarding MRR during ECMM, including various parameters such as electrolyte concentration, voltage, current, duty cycle, and frequency. The increase of every parameter except for pulse frequency led to the in-crease of MRR, whereas current and frequency were found to be the most important parameters. Kumar and Ramanan [16] studied the influence of electrolyte concentration, voltage, and feed rate on MRR and the surface roughness of beryllium copper workpieces. They found that voltage is the most significant parameter for MRR whereas feed rate is the most important parameter for surface roughness.

Apart from the studies which investigate the effect of main process parameters during ECMM, several studies focused on the effect of the tool characteristics. For example, Liu et al. [17] used heavily doped monocrystalline silicon tools with SiO_2 and Si_3N_4 coatings and non-circular cross-section, placed in a rotational head in order to avoid stray current in ECMM. It was shown that acceptable quality and accuracy of micro-grooves and micro-holes can be obtained using these tools. Kozak, Rajurkar, and Makkar [18] conducted a comprehensive work on the effect of micro-ECM tool geometries under various conditions regarding their efficiency in rendering features such as micro-slots, micro-grooves, and micro-holes using non-profile tools. Moreover, they investigated the use of pulsed micro-ECM in order to further improve the accuracy of produced features. Egashira et al. [19] conducted a study on the drilling of micro-holes with a dimension less than 10 μm by means of ultrashort-pulsed micro-ECM under various conditions including different tool shapes, pulse widths and frequencies, voltage, and electrolyte concentration values. The use of a semi-cylindrical tool, under long pulse widths with high frequency, high low-level voltage, and high electrolyte concentration was shown to lead to improved material rate during drilling and better quality of micro-holes. Mithu et al. [20] also performed a study regarding micro-hole creation by ECMM, investigating the effect of several parameters relevant to the tool such as diameter and length, as well as process parameters such as pulse frequency. It was found that larger tools lead to higher MRR, lower machining time, and lower taper angle of the holes, whereas higher tool length decreases MRR and overcut and increases machining time due to short circuits.

Swain, Sundaram, and Rajukar [21] compared the use of coated and uncoated tools for ECMM. They fabricated a nickel-coated W tool and compared it to a conventional W tool, showing that it lead to higher MRR and more stable machining conditions. The ECMM process can produce lower tool wear and is independent of workpiece hardness, allowing higher MRR and precision to be achieved for features such as micro-holes, under the appropriate parameters or techniques such as vibrations of the tool [22]. In this work [22], through the ECMM process, the micro holes were drilled through the surface of the thin workpiece with the required dimensions using a tool holder with a micro-tool vibration system. Ghoshal and Bhattacharyya [23] studied the electrochemical micro-milling of grooves on stainless steel sheets of 35 and 100 μm thickness in a solution of H_2SO_4 of various concentrations. The variations in electrolyte concentration, pulse frequency, and tool feed rate for different types of tools were studied regarding overcut, Ra, and taper. It was shown that overcut was reduced with higher frequency, higher horizontal feed rate, and lower electrolyte concentration, whereas Ra decreases with higher electrolyte concentration and taper angle decreases with a decrease in feed rate.

Wielage et al. [24] investigated the use of carbon fibers as electrode tools for micro-ECM on hard-to-cut materials. They applied this method during the machining of micro-channels and micro-holes and determined the suitable type of fibers which lead to a lower number of defects and can achieve high aspect ratios and smaller overcuts. Chen et al. [25] conducted experiments on ECMM of blisks fabricated by Ti60 material and investigated the effect of several process parameters, such as electrolyte concentration, voltage, pulse frequency, duty cycle, temperature, and anode feed rate on surface roughness. It was found that surface roughness decreases at higher electrolyte concentration, higher frequency, lower duty cycle, and moderate temperature. The correlation between voltage and feed rate was more complex, as lower surface roughness was obtained either with lower or higher voltage and feed rate had a minimal impact on roughness.

Since the electrode has a high influence on various responses in the ECMM process, it is considered that the focus should be set on modifying the tool electrode. Even though various works are available to analyze the effect of various tool electrodes on ECMM performance, only a few works are available relevant to the use of coated electrodes in this process. Hence, in the present analysis, an experimental investigation was made to analyze the effect of a silver-coated copper electrode of ECMM on Incoloy 825 alloy to enhance the process efficacy. The analysis is conducted by statistical methods based on multiple process parameters such as voltage, electrolyte concentration, pulse frequency, and duty factor and indicators such as MRR, overcut, conicity, and circularity. Finally, the superior performance of the silver-coated electrode is determined.

2. Materials and Methods

The experimental arrangement details of the ECMM setup manufactured by SinergyNano Systems (Navi Mumbai, India) in the present study are shown in Figure 1. Electrochemical Micro Machining is a controlled process that involves conductive material removal in the electrolyte chamber where the negative charge is attached to the workpiece, and the positive charge is attached to the tool electrode. The main components used in this ECMM framework are the DC pump, stepper motor, pulsed power supply, and the Acyclic tank size of 200 × 100 × 80 mm. The tank capacity is 1.5 L. The vertical motion of the tool holder is powered by a stepper motor controlled by the microcontroller. The parameters such as supply voltage, feed rate, and duty cycle are given as input for the machine to initiate the hole in the workpiece.

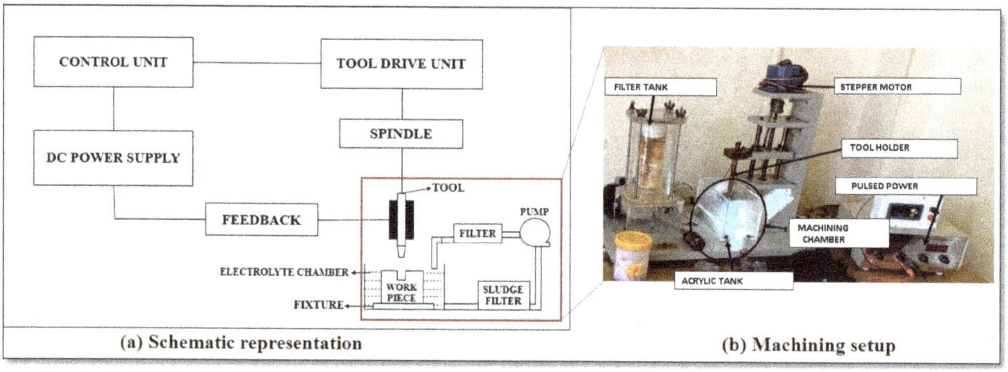

Figure 1. ECMM arrangement in the present study.

2.1. Selection of Workpiece and Electrode

Incoloy 825 is a nickel-iron-chromium-molybdenum-copper superalloy with high amounts of chromium, nickel, copper, and molybdenum that offers excellent corrosion resistance in both mildly oxidized and reducing conditions, as shown in Table 1, where the chemical composition of Incoloy 825 workpiece is reported, measured using the testing method of ASME E3047-16. Since Incoloy 825 has a high nickel content, it is enough to resist chloride ion corrosion cracking stress. The presence of molybdenum and copper enables the material to withstand reducing agents and acids. This material is useful in evaporators such as phosphoric acids, used as the shaft in propellers, for lifting tanks in trucks, heaters such as pickling tanks, calorifiers, hot vessels such as ammonium sulphate, sulphuric acid, and exhaust systems. The Incoloy 825 workpiece specimens with a thickness of 5 mm were processed through a drilling process using the ECMM process.

Table 1. Chemical composition of Incoloy 825.

Elements	Ni	Fe	Cr	Mo	Cu	Ti	C	Mg	Si	S	Al
Composition (%)	42.2	30.89	19.9	2.74	1.637	0.81	0.022	0.71	0.36	0.001	0.116

The tool electrode of the ECMM process should possess thermal conductivity at its peak and high electrical conductivity and corrosion resistance. Based on these criteria, copper (Cu) material of 0.5 mm diameter was selected as a tool electrode. The diameter of the tool electrode has been measured in three positions and the average values were taken for the analysis. The silver-coated copper tool was composed of a copper core covered by concentric silver plating. The conductivity of copper is combined with the bright and polished surface of silver in the electrode. In addition, the silver (Ag) coating provides high corrosion resistance.

2.2. Electroplating Arrangement for Ag-Coated Copper Electrode

Since a thin level of coating is essential for the copper tool electrode, it is difficult to carry out chemical vapor deposition and physical vapor deposition methods. Hence, Electrodeposition was selected for the coating of silver on the copper tool electrode. The bath composition used for the electrodeposition of silver is presented in Table 2. The process was carried out at a current density of 0.3 A/m^2 for 30 min using a DC rectifier. A two-electrode system comprising copper as a cathode and silver mesh as an anode was used.

Prior to electrodeposition, the copper electrode was degreased with acetone and etched with a solution containing 75% H_3PO_4 + 25% H_2SO_4. To enhance the adhesion of silver to the copper tool, a nickel bonding layer was deposited prior to silver electroplating [26]. During the experiment, the bath was mechanically agitated using a magnetic stirrer at 300 RPM. After electrodeposition, the tool was rinsed with distilled water and dried using a hot air oven. The average coating thickness of silver was observed as 8 ± 2 µm.

Table 2. Bath Composition for electrodeposition of silver on copper.

Chemical	Concentration
$AgNO_3$	15 g/L
KCL	15 g/L
$Na_2S_2O_3$	40 g/L
$C_4H_4O_2Na_2$	30 g/L

2.3. Selection of Electrolyte

The electrolyte type has an impact on the machining rate and overcut. The sodium nitrate electrolyte was found to produce a better machining rate and lower overcut. Passivating electrolytes give better precision for machining because they create an oxide layer that provides a better-machined surface, which does not form while using other electrolytes. Sodium nitrate ($NaNO_3$) was selected as an electrolyte for this micromachining process. Since it is a passivating electrolyte with less throwing power, it removes more material, resulting in a high machining rate and precision.

2.4. Selection of the Process Parameters

Since variable parameters are so important in the ECMM method, the following have been chosen as input process parameters for this analysis [27–30]. Applied voltage (AV) is a potential difference between a tool as a cathode and a workpiece as an anode. Based on the machine settings 10 V, 12 V, and 14 V were chosen. The concentration of the electrolyte (CE) is the amount of salt dissolved per liter of water. The electrolyte was used as 20 g, 25 g, and 30 g per liter of distilled water. The frequency (f) was selected for machining as 50 Hz, 60 Hz, and 70 Hz. The duty cycle (DC) indicates the duration at which the energy is applied across the zone. The duty cycle percentage selected for this machining is 33%, 50%, and 66% [31,32]. The micro-holes' quality was analyzed using OPTIV LITE 3020, Hexagon Manufacturing Intelligence, Telford, UK. It was designed to measure the dimensions with an accuracy of 0.001 mm (1 µm) with an inbuilt Charged Coupled Device (CCD) camera machine vision measurement (VMS) system based on the input parameters with the accuracy of 1µm. The optical microscope Olympus BX41M LED Reflected Light Metallurgical Microscope (Olympus, Tokyo, Japan) has also been used to measure the surface performance measures with the magnification range of 50×–500×. The calibration has been done using the Metalplus inbuilt software.

MRR, overcut, conicity, and circularity were chosen as response measures that must be carried out in a theoretical and graphical manner for the experiment to be efficient [33,34]. MRR is the amount of material removed per minute when turning, cutting, milling, grinding, or honing metal. The actual and theoretical MRR can be computed as Equations (1) and (2).

$$MRR_{Actual} = \frac{W1 - W2}{\rho \cdot T} \; [mm^3/min] \tag{1}$$

where W1 = weight of the workpiece before machining in gr; W2 = weight of the workpiece after machining in g; T = time for machining in min; ρ = Density of material in g/cc.

$$MRR_{Theoretical} = \frac{I \cdot A}{\rho \cdot V \cdot F \cdot 60} \; [mm^3/min] \tag{2}$$

where I = Current in A; A = atomic weight in amu; V = valence; F = Faraday's constant equal to 96,500 C mol^{-1}; A/V = e (equivalent weight). The diameter deviation from the electrode tool and the machined hole is called an overcut (OC) and can be calculated using Equation (3), as follows [35]:

$$OC = \frac{Di - Dt}{2} \ [\mu m] \quad (3)$$

where, Dt = external tool diameter in mm; Di = average diameter of the workpiece in mm. Overcut is an important parameter when assessing micro-hole quality [36]. It is to be noted that in the relevant literature, researchers prefer either to calculate the overcut regarding diameter [37] or the radial overcut [35,38–41], which is calculated by Equation (3). In any case, the trends observed in the analysis will be similar.

The degree of straightness of the through-hole that is machined is called conicity (CY) and it can be calculated using Equation (4) [42–44]:

$$CY = \frac{\text{Diameter at entry} - \text{Diameter at exit}}{2 \cdot \text{hole thickness}} \quad (4)$$

This quantity, based on its definition by Equation (4), is similar to the taper rate or the tangent of taper angle, calculated in various machining processes and should be minimum for straighter holes [43–46]. Conicity is one of the important parameters regarding the shape accuracy of produced features, especially holes [44,46–49] as under various conditions the produced holes are usually cone-shaped instead of cylindrical. This parameter can also indicate the morphology of the holes, as when the entry diameter is lower than the exit diameter, the negative value of CY indicates a hole with reversed conicity [42]. In general, conicity can be considered a major goal during the fabrication of microelectronic parts, thus it is fundamental to determine the optimum conditions for minimum conicity of micro-holes [45].

The number of experiments was estimated using the Taguchi approach and the experimental design was determined using the L_9 orthogonal array (OA). Since the study dealt with four parameters and three levels L_9 OA was fixed as shown in Table 3. It is to be noted that, every measurement was repeated three times and the average values were reported.

Table 3. Design of Experiment using Coated and Uncoated Tool Electrode.

Exp No.	Applied Voltage [V]	Electrolyte Concentration [g/L]	Frequency [Hz]	Duty Cycle [%]
1	10	20	50	33
2	10	25	60	50
3	10	30	70	66
4	12	20	60	66
5	12	25	70	33
6	12	30	50	50
7	14	20	70	50
8	14	25	50	66
9	14	30	60	33

3. Results and Discussion

The Incoloy 825 workpiece specimens with a thickness of 5 mm were processed through a drilling process using the ECMM process. The experimental values obtained under the L_9 approach were discussed afterward and analyzed; a comparison of the performance of Ag-coated and uncoated copper electrodes was also performed.

3.1. Microstructure Analysis of Ag-Coated and Uncoated Copper Tool Electrode

The surface morphology of the electrodes in the ECMM process was observed using the SIPCON vision measuring system. The surfaces have been polished before the investigation as per ASTM standard. Figure 2 shows the surface morphology of uncoated and silver-coated electrodes before machining in the ECMM process.

Figure 2. Morphology of copper electrode before machining (**a**) Uncoated (**b**) Silver Coated.

It was observed that the cracks on the tool surface were reduced with the coated electrodes. This has resulted in the formation of a smooth surface with the Ag-coated electrode as seen in the surface morphology images. The coating of silver on the copper tool increases the electrical conductivity of the tool electrode when compared to the uncoated Copper Tool Electrode. Figure 3 displays the surface morphology of uncoated and silver-coated copper tool electrodes after machining in the ECMM process. From analyzing the surface morphology images of tool electrodes after the machining process, more pits and pores could be observed due to the interchange of the ions from the tool to the workpiece.

Figure 3. Surface Morphology of copper electrode after machining (**a**) Uncoated (**b**) Silver Coated.

3.2. Electrical Conductivity Analysis of Ag-Coated and Uncoated Copper Tool Electrode

The resistance of the copper tool electrode was measured before and after the coating process using a V-I graph with the help of a desktop Multimeter which has a range of 1micro ohm to 100 ohms. Table 4 shows that the electrical conductivity (σ) of the Cu tool electrode was increased with the silver coating due to the higher conductivity of Ag than Cu. The slope graph was generated using the run values obtained in resistance. Then, the slope equation, resistance, and area of the tool are substituted in the conductivity Equation (5).

$$\rho_{res} = \frac{R \cdot A}{L} \; [\Omega \text{ mm}] \tag{5}$$

where, ρ_{res} = resistivity in Ω mm; R = resistance; $A = \pi r^2$ the area in mm^2; r the radius in mm; L the length of wire in mm.

$$\sigma = \frac{1}{\rho_{res}} \left[\Omega^{-1}/\text{mm}\right] \tag{6}$$

Table 4. Conductivity before and after machining.

Electrode	Electrical Conductivity [Ω^{-1}/mm]	
	Before Machining	After Machining
Uncoated	31.41	2.93
Ag coated	34.58	3.48

It was observed that the conductivity of the copper tool is the least among the two. The movement of atoms in the crystal lattice causes the ductility and stiffness of the copper tool electrode to increase, lowering the conductivity. The decrease in the electrical conductivity was observed after the machining process. The electrical resistance of the tool increases as the temperature of the tool electrode increases during machining, causing the atoms to vibrate and collide with moving electrons. The hindrance of moving electrons reduces the conductivity of the tool electrode.

3.3. *Influence of Ag-Coated Electrode and Process Parameters on Actual MRR*

From Table 5, it was observed that the MRR of the Ag-coated copper tool was higher than the MRR of the uncoated copper tool in most cases. Furthermore, except for a few cases, the results obtained by different electrodes are statistically different, as can be clearly observed in Figure 4. The maximum MRR was obtained under a duty factor value of 33% at the high-value voltage and electrolyte concentration as well as a moderate value of pulse frequency, namely 14 V, 30 g/L, and 60 Hz. From the Main effects plots of Figure 4, it can be seen that different trends are obtained when the uncoated and coated electrode is used. Regarding voltage, the use of silver-coated electrode leads to a considerable increase of actual MRR, especially for voltage values over 12 V, for which the results obtained by different electrodes are statistically different as can be seen by the error bars in Figure 4, compared to the use of the uncoated tool, although the trend is common. Regarding electrolyte concentration, the use of a silver-coated tool not only increases the MRR with increased electrolyte concentration, but also changes the trend considerably, with an increasing trend observed for the silver-coated tool while a slightly decreasing trend was observed for the uncoated tool. The pulse frequency affects MRR in a different way when silver-coated and uncoated electrodes were used, as MRR was minimized at a moderate frequency value for the silver-coated electrode, whereas it was maximized for the same frequency value for the uncoated tool. Finally, the MRR obtained with the silver-coated tool is higher for every duty factor value used and the trend is similar for both electrodes, with MRR being minimum at moderate duty factor values.

Table 5. Response table on actual MRR.

S. No	V [V]	CE [g/L]	f [Hz]	DC [%]	Actual MRR [mm^3/min]	
					Uncoated	Ag Coated
1	10	20	50	33	0.00063	0.000499
2	10	25	60	50	0.000805	0.000592
3	10	30	70	66	0.000505	0.00087
4	12	20	60	66	0.000858	0.000415
5	12	25	70	33	0.000529	0.00076
6	12	30	50	50	0.000529	0.000809
7	14	20	70	50	0.000569	0.000584
8	14	25	50	66	0.000671	0.001029
9	14	30	60	33	0.000923	0.001051

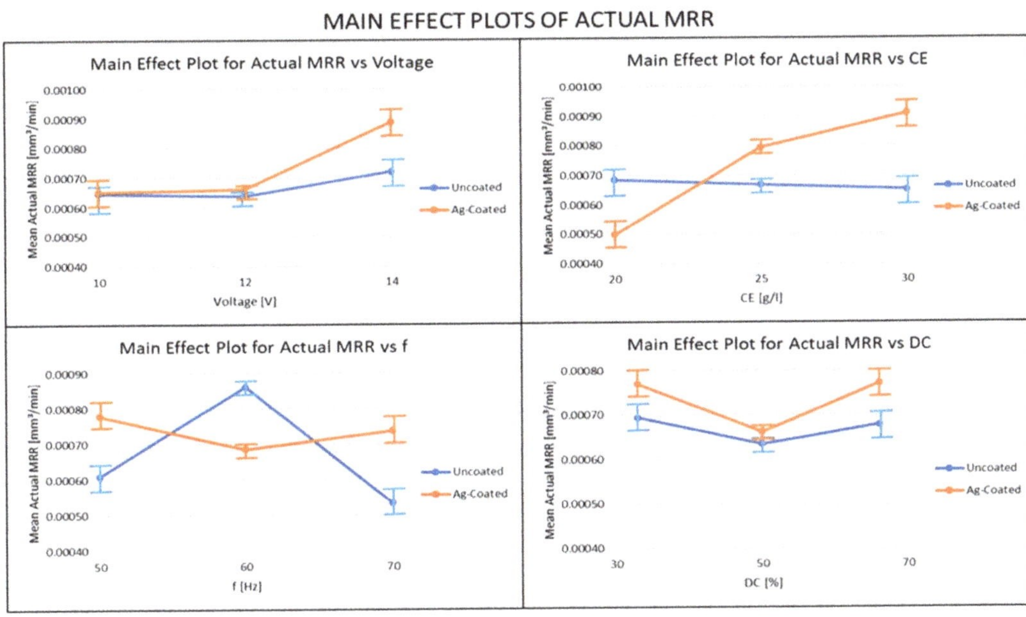

Figure 4. Main Effects Plots of the actual MRR.

Based on the statistical analysis of the Taguchi DoE using ANOVA, it is deduced that for the actual MRR for the uncoated electrode, the most significant parameter is the frequency followed by the voltage, while the duty factor and the concentration of the electrolyte have the less impact on the actual MRR. However, for the Ag-coated electrode, the most significant parameter is the concentration of the electrolyte, followed by the voltage, whilst the duty factor and the frequency are the less important parameters.

Electrolyte concentration was proven to be the most influencing parameter for the coated copper tool electrode due to fine grain structure compared to uncoated electrodes. The anodic dissolution is proportional to the conductivity of the machining zone. Since the Ag has higher electrical conductivity than the Cu electrode, the MRR obtained was higher. A higher electrolyte concentration contributes to more ions in the electrolytic medium, resulting in a faster dissolution rate. As the number of ions in the medium increases, there is an increased density of current that leads to a high amount of material removal if led to the higher dissolution of electrolytes and particles removed from the workpiece, thereby increasing the MRR.

Moreover, the higher voltage leads also to a higher MRR due to the increase of the rate of ionization and the enhancement of dissolution of the workpiece material. However, it can also affect the electrolyte properties due to an increased amount of heat produced at higher voltage, something that contributes to the higher removal rate. This effect was greater for the Ag-coated electrode due to its larger conductivity. The pulse frequency is affecting MRR both directly and indirectly, as it can affect the number of pulses per unit of time and thus, the amount of material removed at a specific time period. On the other hand, it can affect electrolyte temperature and flow conditions, as well as electrode wear, which have contradicting effects on MRR. That is the reason why it leads to a larger variation of MRR for the uncoated electrode, which is more prone to wear, than for the Ag-coated one. Finally, the duty cycle is directly related to the ratio of pulse on time and total pulse duration; thus, increased pulse on time leads to more material removal but also increases electrode wear so the MRR cannot be considerably increased.

These results are considerably important, as MRR is related to a crucial goal of the process, i.e., productivity, which is essential in industrial applications, such as hole drilling, apart from obtaining good surface quality and appropriate dimensional accuracy. In the case of ECMM hole drilling, it can be suggested that, for obtaining high MRR, it is more favorable to use a coated electrode at high voltage and electrolyte concentration, relatively low frequency and high duty cycle values.

3.4. Influence of Ag-Coated Electrode and Process Parameters on Theoretical MRR

From Table 6, it was shown that the larger theoretical MRR was obtained for a voltage value of 14 V, electrolyte concentration of 30%, pulse frequency of 60 Hz, and duty factor value of 33%, the same conditions where the actual MRR was the highest. A higher electrolyte concentration shows a considerable amount of ions in the electrolytic medium, resulting in a faster dissolution rate. When the number of ions in the medium rises, current density could result in greater material removal. It was found that the theoretical MRR of the Ag-coated copper tool was better than the MRR obtained by the uncoated copper tool in most cases. The fine grain structure of coated tool electrode provides higher dissolution and removal of particles from the workpiece, leading to an increase in the material removal rate. Although for the theoretical MRR it is anticipated that the results between cases where the two different electrodes were used will be more similar than the results for the actual MRR, something that is justified by the error bars of Figure 5, especially for low to moderate voltages, low to moderate frequencies and duty cycles, in most cases where the results have statistically significant difference, the MRR obtained with the coated electrode is higher. Since the Ag has higher electrical conductivity than Cu electrode, the MRR was higher.

Table 6. Response table on theoretical MRR.

S. No	V [V]	CE [g/L]	f [Hz]	DC [%]	Theoretical MRR [mm^3/min]	
					Uncoated	Ag Coated
1	10	20	50	33	0.00098	0.000804
2	10	25	60	50	0.000921	0.000882
3	10	30	70	66	0.000804	0.001
4	12	20	60	66	0.001078	0.000843
5	12	25	70	33	0.000804	0.001019
6	12	30	50	50	0.001058	0.001137
7	14	20	70	50	0.000784	0.000784
8	14	25	50	66	0.00096	0.001117
9	14	30	60	33	0.001215	0.001352

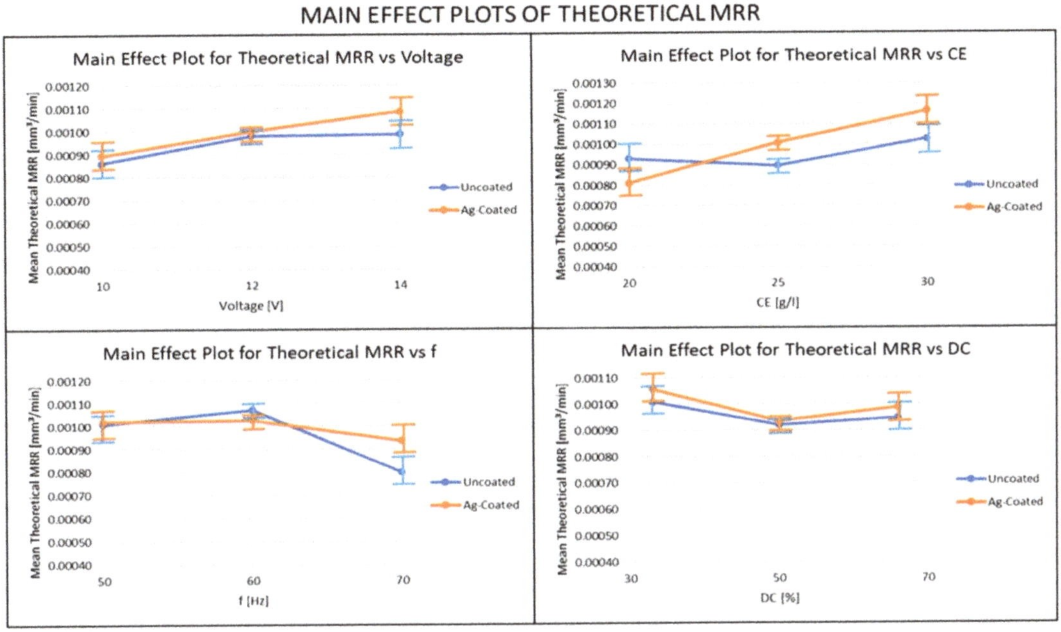

Figure 5. Main Effects Plots of the theoretical MRR.

From the Main effects plots of Figure 5 it can be seen that theoretical MRR has different trends than Actual MRR and different trends are also obtained when the uncoated and coated electrode is used. Regarding voltage, the use of a silver-coated electrode leads to a slight increase in the actual MRR, especially for voltage values over 12 V, compared to the use of the uncoated tool, for which MRR becomes almost constant after 12 V. Regarding electrolyte concentration, the use of a silver-coated tool not only increases the MRR with increased electrolyte concentrations, but also changes the trend considerably, with an increasing trend observed for the silver-coated tool while a slightly decreasing and increasing trend was observed for the uncoated tool. The pulse frequency affects MRR in a different way when silver-coated and uncoated electrodes were used, as MRR slightly reduced for increasing pulse frequency when the silver-coated electrode was used, whereas it was slightly increased and then considerably decreased for the uncoated tool. Finally, the MRR obtained with the silver-coated tool is higher for every duty factor value used and the trend is similar for both electrodes, with MRR being minimum at moderate duty factor values.

Again, based on ANOVA results, the significance of the parameters is not the same for the uncoated and the Ag-coated electrode. More specifically, for the uncoated electrode, the most significant machining parameter is the frequency followed by the concentration of the electrolyte and the voltage while the duty factor is the less important parameter. On the other hand, for the Ag-coated electrode, the concentration of the electrolyte is the most important parameter followed by the voltage, while the duty factor and the frequency are again the parameters with the less impact.

3.5. *Influence of Ag-Coated Electrode and Process Parameters on Overcut*

From the Main effects plots of Figure 6, it can be seen that different trends regarding overcut are obtained when uncoated and the coated electrode is used. Regarding voltage, the use of silver-coated electrode leads generally to a lower overcut, and an increase of voltage results in an increase of overcut, especially for voltage values around 12 V, compared to the use of the uncoated tool, for which overcut first increases and then decreases slightly after 12 V. Regarding electrolyte concentration, opposite trends are observed when silver-coated and uncoated electrodes are used. More specifically, the use of a silver-coated tool leads to a maximum overcut at a moderate level of electrolyte concentration, while the minimum overcut at the same level is observed for the uncoated tool, whereas overcut is generally lower for the coated tool. The pulse frequency affects overcut in a different way when silver-coated and uncoated electrodes were used, as overcut remains almost constant, with a small decreasing trend for increasing pulse frequency when the silver-coated electrode was used, whereas it was considerably reduced at higher frequency values for the uncoated tool. Finally, the overcut values obtained with the silver-coated tool are reduced for duty factor values over 50% whereas overcut is clearly increased when the duty factor is increased for the uncoated electrode.

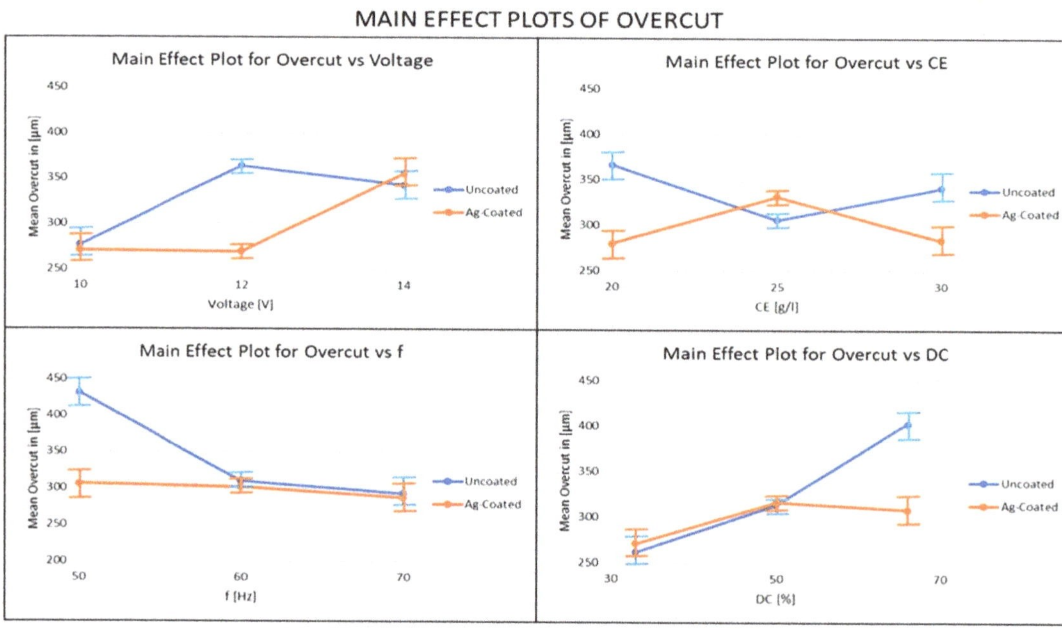

Figure 6. Main Effects Plots of the overcut.

The micro-holes' quality was analyzed using a machine vision measurement (VMS) system based on the input parameters as shown in Figure 7. The micro-hole quality was designated as better, good, and poor for the applied process parameters based on the degree of deviation from the appropriate hole diameter. From Table 7 it can be seen that the lowest overcut was obtained at 10 V, 25 g/L electrolyte concentration, 60 Hz pulse frequency and duty factor of 50% for the uncoated electrode, and at 10 V, 20 g/L electrolyte concentration, 50 Hz pulse frequency and duty factor of 33% for the silver-coated electrode, with the overcut value for the uncoated electrode being slightly lower than that of the coated one. However, in general, the overcut values obtained with the uncoated electrode were higher, with the highest overcut obtained at 14 V, 25 g/L electrolyte concentration, 50 Hz pulse frequency, and duty factor of 66%.

Regarding the impact of each parameter on the OC based on ANOVA, interestingly, an almost opposite trend is observed for the uncoated and the Ag-coated electrode. Namely, the parameters for the uncoated electrode in descending order of significance are the duty factor, the frequency, the voltage, and the concentration of the electrolyte, while for the Ag-coated electrode are the voltage, the concentration of the electrolyte, the duty factor, and the frequency.

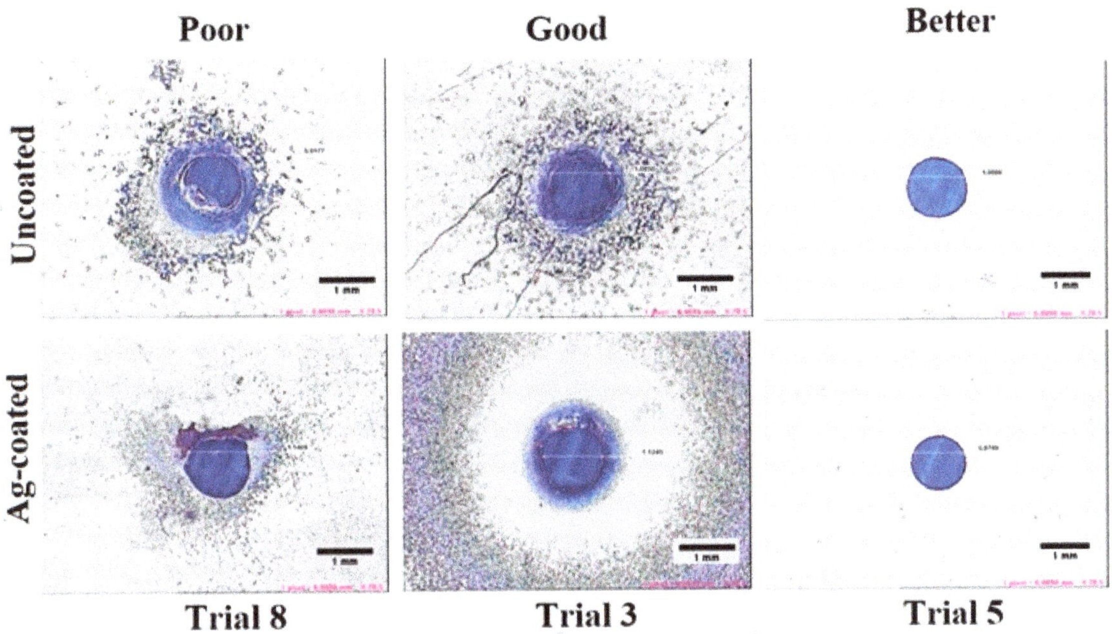

Figure 7. VMS Image for Overcut.

Table 7. Response table of the overcut.

S. No	V [V]	CE [g/L]	f [Hz]	DC [%]	Overcut [μm]	
					Uncoated	Ag Coated
1	10	20	50	33	284.5	232.4
2	10	25	60	50	223.3	325.5
3	10	30	70	66	330.0	255.2
4	12	20	60	66	429.9	264.4
5	12	25	70	33	246.3	264.2
6	12	30	50	50	415.6	280.4
7	14	20	70	50	302.6	343.4
8	14	25	50	66	448.4	405.7
9	14	30	60	33	277.3	316.2

The lower overcut for the coated electrode can be attributed to the fine grain structure of the silver-coated copper tool electrode, which limits electron movement to the tool's edge, increasing the current available at the frontal gap between the workpiece and the tool. The increase in voltage leads to less localized current flux and higher stray current so that more material is removed from the workpiece [35]. Moreover, increased electrolyte concentration leads to a higher concentration of ions and reduces the localization of material removal, thus producing higher overcuts up to a point [38], especially for the Ag-coated tool, as for the uncoated tool, the higher wear can cause a variation of this behavior. The higher duty factor, caused by a higher pulse on time leads to a higher amount of removed material, thus higher overcut is obtained. The fact that the duty factor value of 50% is critical can be attributed to the material removal time being equivalent to the flushing time at 50% of the service cycle. Finally, with an increase in frequency value, the overcut decreases due to a higher pulse generated per unit time something that leads to an improvement in localization while stray current decreases, resulting in a low overcut.

Lower overcut is also an important goal of the process, apart from higher productivity, as dimensional accuracy, especially for micro-holes, is necessary for various high-end applications, e.g., microelectronics. From the findings of this work, it is suggested that a coated electrode and low voltage, electrolyte concentration, and duty cycle along with high-frequency values, lead to the minimization of overcut.

3.6. Influence of Ag-Coated Electrode and Process Parameters on Conicity

As shown in Table 8, the lowest value for conicity for the uncoated electrode is obtained with a voltage value of 14 V, electrolyte concentration of 25 g/L, pulse frequency of 50 Hz, and duty cycle value of 66% whereas the lowest value of conicity for the coated electrode is obtained with a voltage value of 12 V, electrolyte concentration of 25 g/L, pulse frequency of 70 Hz and duty cycle value of 33%. On the other hand, the maximum value for conicity for the uncoated electrode is obtained with a voltage value of 12 V, electrolyte concentration of 25 g/L, pulse frequency of 70 Hz, and duty cycle value of 33%, whereas for the coated electrode, the higher value is obtained for voltage value of 10 V, electrolyte concentration of 30 g/L, pulse frequency of 70 Hz, and duty cycle value of 66%. Although the minimum conicity value was obtained by the uncoated electrode, in most cases, the conicity is lower when the coated electrode is used. Moreover, for the uncoated electrode, conicity values fluctuate considerably higher than the conicity values for the coated electrode.

Table 8. Response table on conicity.

S. No	V [V]	CE [g/L]	f [Hz]	DC [%]	Conicity [Degree]	
					Uncoated	Ag Coated
1	10	20	50	33	11.55	4.98
2	10	25	60	50	8.27	3.06
3	10	30	70	66	0.80	5.60
4	12	20	60	66	0.86	3.92
5	12	25	70	33	27.93	1.60
6	12	30	50	50	6.67	2.73
7	14	20	70	50	1.64	3.25
8	14	25	50	66	0.07	1.88
9	14	30	60	33	5.86	1.64

From the Main effects plots of Figure 8, it can be seen that different trends regarding conicity are obtained when uncoated and coated electrodes are used. Regarding voltage, the use of silver-coated electrodes leads generally to lower conicity, and an increase in voltage results in a slight decrease of conicity, compared to the use of an uncoated tool, for which conicity first increases and then decreases considerably after 12 V. Regarding electrolyte concentration, opposite trends are observed when silver-coated and uncoated electrodes are used. More specifically, the use of a silver-coated tool leads to a minimum conicity at a moderate level of electrolyte concentration, while the maximum conicity at the same level is observed for the uncoated tool, whereas conicity is generally lower for the coated tool. The pulse frequency affects conicity to a different extent when silver-coated and uncoated electrodes were used, as conicity remains almost constant, with a small increasing trend for increasing pulse frequency when the silver-coated electrode was used, whereas it was slightly reduced and then increased at higher frequency values for the uncoated tool. Finally, the overcut values obtained with the silver-coated tool are almost constant whereas overcut is clearly decreased when the duty factor is increased for the uncoated electrode.

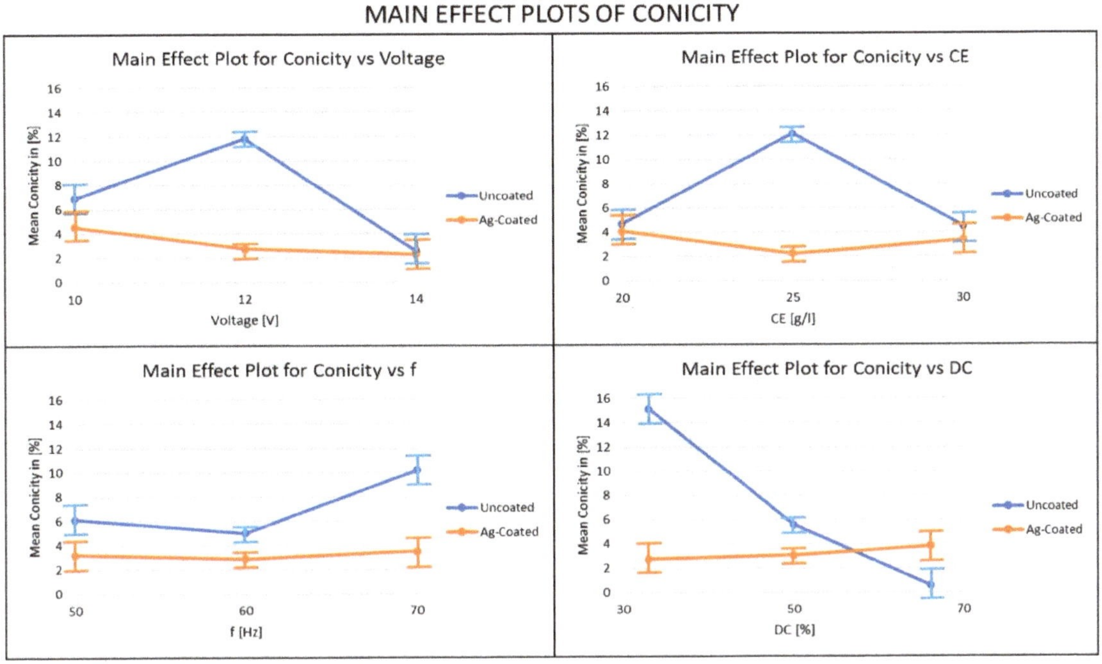

Figure 8. Main Effects Plots of the conicity.

For the conicity based on ANOVA results, it is deduced that for the uncoated electrode, the most important machining parameter is the duty factor, followed by the voltage, the frequency, and the concentration of the electrolyte. However, when the Ag-coated electrode is employed, the voltage becomes the most important parameter, followed by the concentration of the electrolyte, the duty factor, and the frequency.

The increase in voltage leads to higher energy, higher wear, and variations in the electrolyte flow, thus it increases material removal and conicity, especially for the uncoated electrode. However, in the case of the Ag-coated electrode, the coating leads to lower deformation of the tool and finally, the conicity is barely affected. The increase of electrolyte concentration leads generally to higher current density and material removal, thus can increase conicity up to a point for the uncoated electrode, whereas further increase can lead to a different behavior due to increased wear and variation of electrode shape. On the other hand, the use of coating prevents undesirable phenomena and leads to an almost constant variation of conicity with respect to the electrolyte concentration. Moreover, the higher frequency can affect material removal to a larger extent and thus conicity is increased, especially for the uncoated electrode.

In order to achieve low conicity, the use of a coated electrode is recommended, along with high voltage, moderate electrolyte concentration, low frequency, and low duty cycle values.

3.7. SEM Analysis of the Workpiece Specimen and Tool Electrode

Figure 9 shows scanning electron microscope (SEM: JEOL JAPAN, Jeol-6490 JED-2300) images of the machined workpieces with uncoated and Ag-coated electrodes. It was used to examine the surface topography of the electrochemically micro-machined Incoloy 825 workpiece.

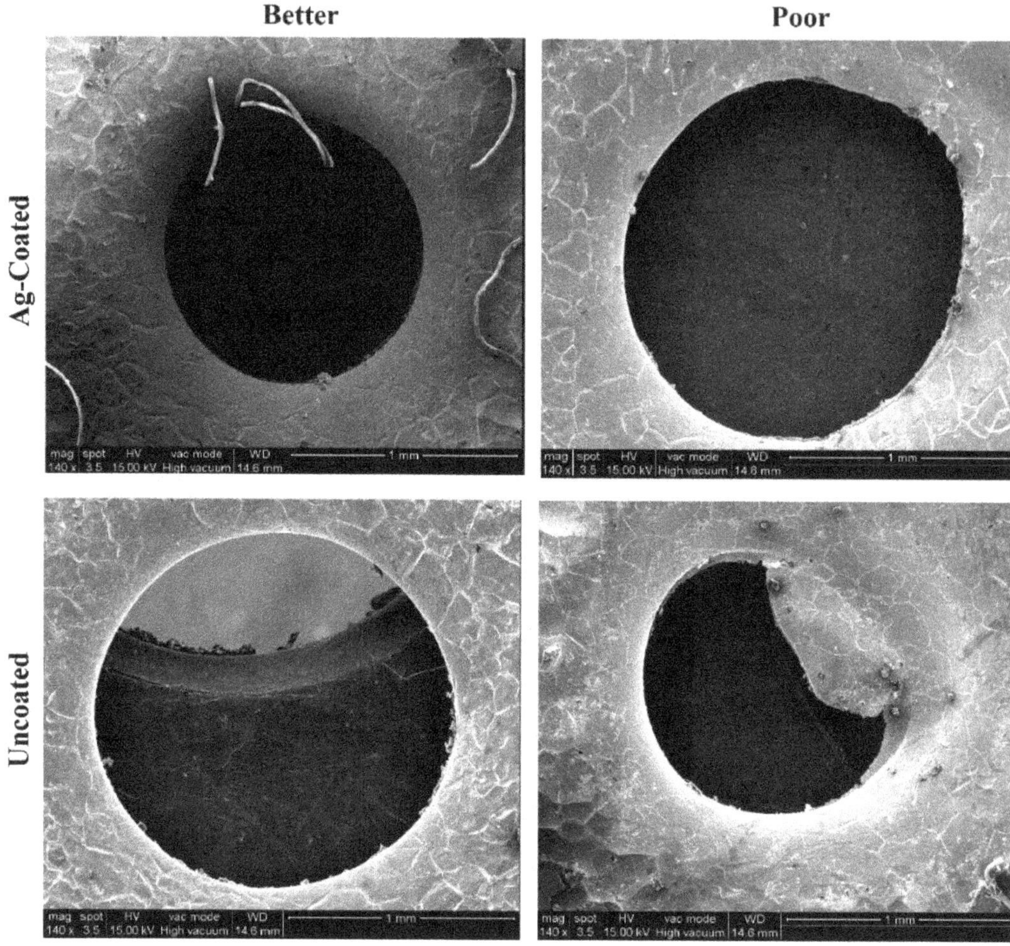

Figure 9. SEM Image of Machined Coated and Uncoated Workpiece.

According to the investigation of all images considered, the machining accuracy was higher when the $NaNO_3$ electrolyte reacted with the silver-coated copper tool electrode on Incoloy 825 workpiece. From these images, qualitative observations regarding the circularity error can be conducted. The circularity error is defined as the extent of the deviation of a machined hole from the perfect circle and can be directly affected by the process parameters [37]. Ag-coated electrodes could produce a smoother shape, crack-free with lesser circularity error than uncoated electrodes. The uncoated tool lacks prescribed circularity due to the formed cracks. It was indicated that the machining process with low input parameters could reduce the risk of sparking and stray current. It was found the effect of sparking and stray current tends to be evident in the machined workpiece, leading to some irregularities in the edges of the machined surface at high input parameters (14 V, 70 Hz, and 25 g/L).

Figure 10 defines the surface examination of machined silver-coated copper tool electrodes and uncoated copper tool electrodes using SEM. From the analysis, it was observed that uncoated copper tool electrode tends to have more cracks and corroded regions. These cracks and corroded regions represent the presence of carbon content in electrolyte dissolution which has reflected its effect on the tool during the ECMM process.

However, when silver-coated copper tool electrode is observed, it tends to exhibit fewer cracks and corroded regions than the uncoated copper tool electrode. It clearly shows that the silver coating acts as a corrosion barrier when combined with carbon-based electrolyte dissolution, resulting in less carbon formation on the workpiece during machining.

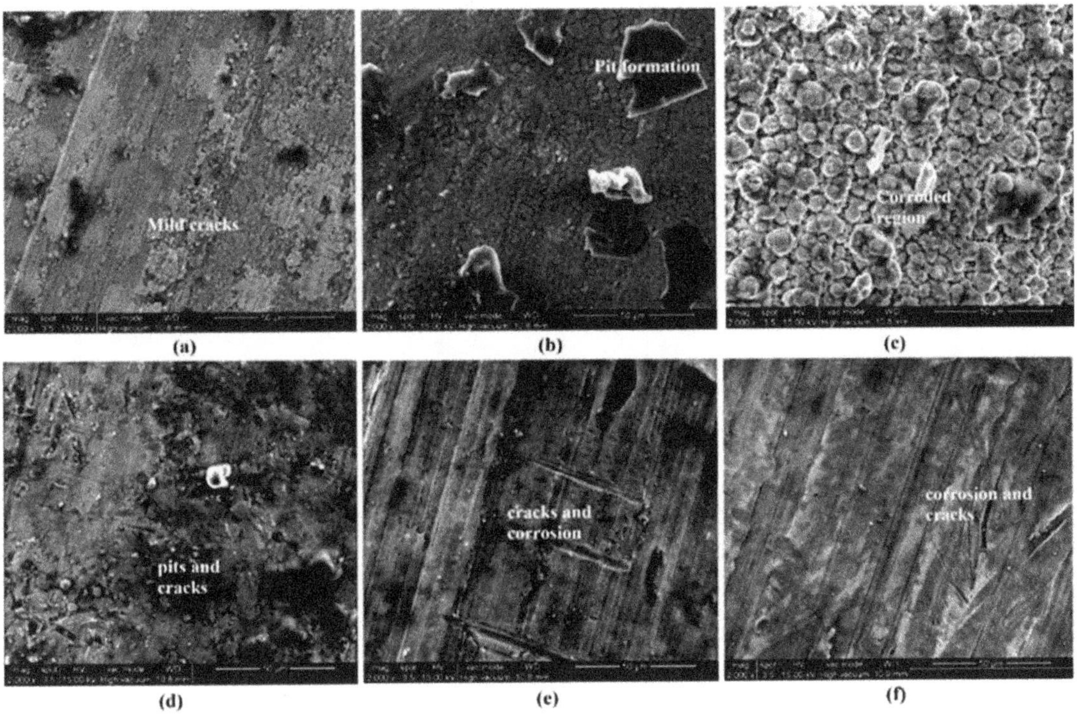

Figure 10. SEM Image of Machined Tool Electrodes (**a**) Ag-coated under trial 5 (**b**) Ag-coated trial 3 (**c**) Ag-coated trial 8 (**d**) Uncoated under trial 5 (**e**) Uncoated trial 3 (**f**) Uncoated trial 8.

From the observations of the SEM images, the silver-coated copper tool electrode which is used to work on input parameters of 14 V, 50 Hz, 25 g/L, and duty cycle of 66%, was found to have more uneven localized deposition on the tool surface during the study. It is due to the reaction of a coated tool under relatively harsh process conditions, which causes stray current corrosion in the tool electrode. On the other hand, under a favorable combination of low input conditions and coated tool, which resulted in less stray current and corrosion in the tool electrode, the forming of irregular localized deposition on the tool surface is minimized, and even the cracks will be seen less when working with low and optimum input parameter values.

3.8. EDS Analysis of the Workpiece Specimen and Tool Electrode

In this subsection, the EDS results regarding the machined workpieces and tool electrodes will be analyzed. It is to be noted that the measurements were conducted on selected points on the side walls of the produced hole and the bottom surface of the tool electrode, respectively. In Figures 11 and 12, the results regarding the workpiece under the conditions of trials 5 and 8 are presented, both when the uncoated and Ag-coated tools were used. Moreover, in Figures 13 and 14, the results regarding the uncoated and coated electrode under the conditions of trials 5, 3, and 8 are presented.

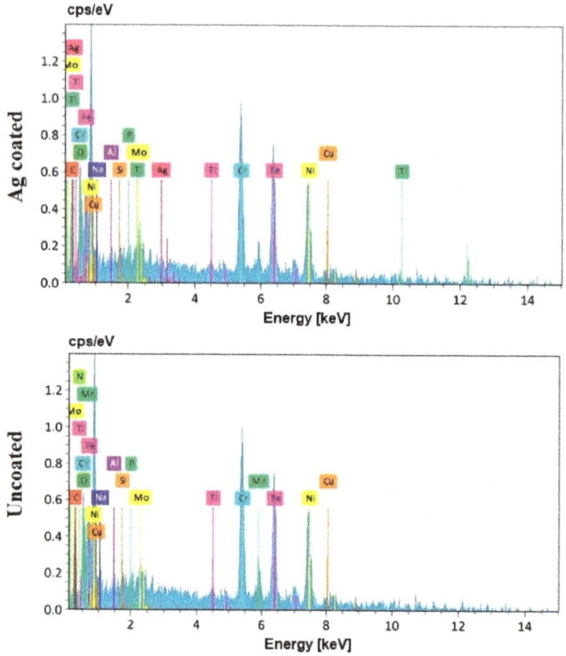

Element	At. No.	Mass [%]	Mass Norm. [%]	Atom [%]	abs. error [%] (1 sigma)	rel. error [%] (1 sigma)
Ni	28	34.99	35.94	26.44	1.89	5.39
Fe	26	24.97	25.65	19.83	1.27	5.10
Cr	24	20.47	21.02	17.46	0.97	4.73
C	6	4.70	4.82	17.35	2.60	55.43
O	8	3.67	3.77	10.17	1.49	40.73
Cu	29	2.95	3.03	2.06	0.51	17.39
Na	11	1.94	1.99	3.75	0.34	17.51
Mo	42	1.79	1.84	0.83	0.21	11.92
Ti	22	0.91	0.93	0.84	0.16	18.08
Si	14	0.41	0.42	0.64	0.10	24.29
Al	13	0.27	0.28	0.44	0.09	32.46
Tl	81	0.21	0.21	0.04	0.09	42.46
P	15	0.11	0.11	0.15	0.06	55.20
		97.36	100.00	100.00		

Element	At. No.	Mass [%]	Mass Norm. [%]	Atom [%]	abs. error [%] (1 sigma)	rel. error [%] (1 sigma)
Ni	28	33.75	33.52	21.50	1.82	5.39
Fe	26	24.05	23.89	16.10	1.23	5.10
Cr	24	19.86	19.73	14.29	0.94	4.72
C	6	9.99	9.92	31.10	3.97	39.74
O	8	3.86	3.83	9.02	1.56	40.31
Cu	29	2.85	2.83	1.68	0.50	17.42
Na	11	1.90	1.89	3.10	0.33	17.42
Mo	42	1.73	1.71	0.67	0.20	11.75
Mn	25	1.00	0.99	0.68	0.19	19.08
Ti	22	0.87	0.86	0.68	0.16	18.20
Si	14	0.39	0.39	0.52	0.10	24.50
Al	13	0.26	0.26	0.36	0.09	32.86
P	15	0.11	0.11	0.14	0.06	51.92
N	7	0.06	0.06	0.17	0.26	408.01
		100.68	100.00	100.00		

Figure 11. EDS Image of Machined Workpiece with Ag-Coated and Uncoated tool under trial 5.

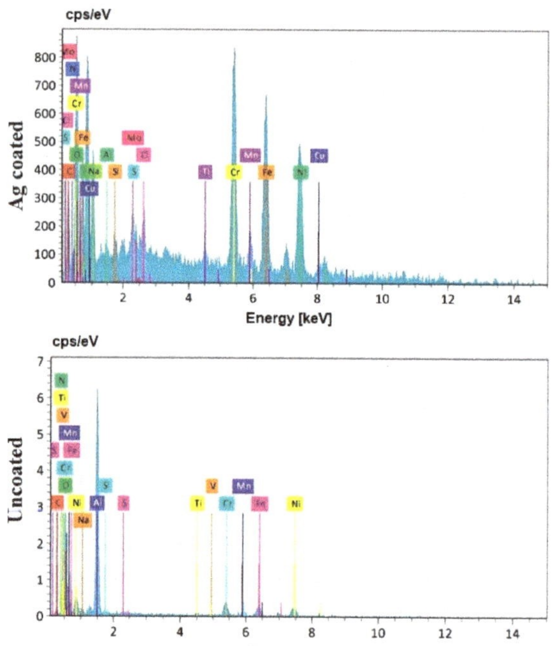

Element	At. No.	Mass [%]	Mass Norm. [%]	Atom [%]	abs. error [%] (1 sigma)	rel. error [%] (1 sigma)
Ni	28	29.76	28.54	17.66	1.31	4.39
Fe	26	23.89	22.91	14.89	0.98	4.10
Cr	24	20.49	19.65	13.72	0.79	3.87
O	8	7.76	7.44	16.89	1.86	23.92
C	6	7.30	7.00	21.18	2.44	33.36
Na	11	4.79	4.59	7.26	0.46	9.64
Cu	29	2.68	2.57	1.47	0.33	12.24
Mo	42	2.42	2.32	0.88	0.20	8.07
Cl	17	1.15	1.11	1.13	0.11	9.66
Ti	22	1.10	1.06	0.80	0.13	11.46
Mn	25	1.00	0.96	0.63	0.14	13.90
N	7	0.88	0.85	2.19	0.65	73.83
Si	14	0.54	0.52	0.67	0.09	15.90
Al	13	0.35	0.34	0.45	0.08	22.00
S	16	0.16	0.15	0.17	0.05	32.86
		104.28	100.00	100.00		

Element	At. No.	Mass [%]	Mass Norm. [%]	Atom [%]	abs. error [%] (1 sigma)	rel. error [%] (1 sigma)
Al	13	48.93	44.16	48.17	2.48	5.07
Ni	28	18.87	17.03	8.54	1.24	6.60
Fe	26	12.65	11.42	6.02	0.82	6.48
Cr	24	12.24	11.04	6.25	0.70	5.68
C	6	8.94	8.07	19.77	4.56	51.01
O	8	4.28	3.86	7.11	1.76	41.23
Si	14	2.51	2.26	2.37	0.25	9.90
Ti	22	0.68	0.61	0.38	0.14	20.87
Na	11	0.65	0.59	0.75	0.15	23.12
Mn	25	0.65	0.58	0.31	0.16	24.71
S	16	0.41	0.37	0.34	0.09	22.77
V	23	0.00	0.00	0.00	0.00	0.86
N	7	0.00	0.00	0.00	0.00	10.00
		110.80	100.00	100.00		

Figure 12. EDS Image of Machined Workpiece with Ag-Coated and Uncoated tools under trial 8.

(a)

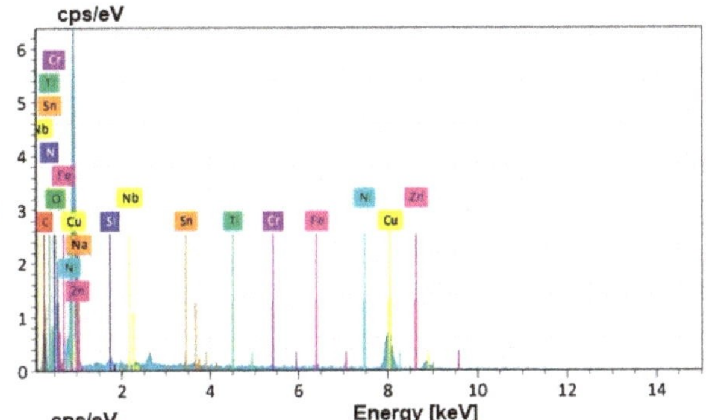

Element	At. No.	Mass [%]	Mass Norm. [%]	Atom [%]	abs. error [%] (1 sigma)	rel. error [%] (1 sigma)
Cu	29	61.64	53.05	25.21	2.59	4.21
C	6	22.26	19.16	48.17	5.12	22.99
Hf	72	15.62	13.44	2.27	1.13	7.25
O	8	13.73	11.82	22.31	2.98	21.73
Cl	17	1.37	1.18	1.01	0.12	9.08
Fe	26	0.90	0.78	0.42	0.14	15.07
Si	14	0.66	0.57	0.61	0.10	15.10
		116.17	**100.00**	**100.00**		

(b)

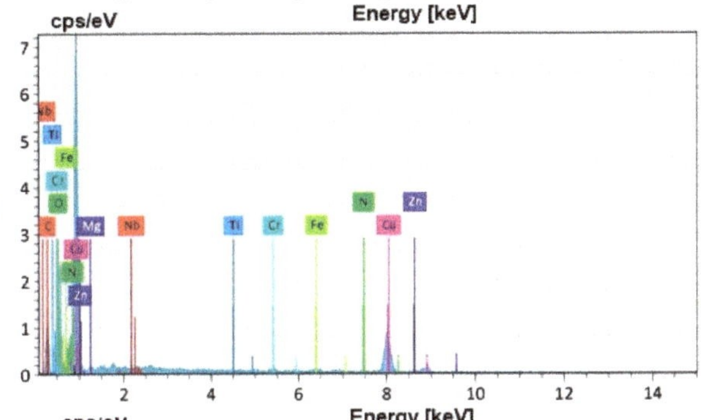

Element	At. No.	Mass [%]	Mass Norm. [%]	Atom [%]	abs. error [%] (1 sigma)	rel. error [%] (1 sigma)
Cu	29	63.86	66.19	31.06	2.43	3.81
C	6	19.62	20.33	50.47	4.15	21.17
O	8	7.89	8.17	15.23	1.73	21.94
Zn	30	2.42	2.51	1.14	0.31	12.62
Cr	24	0.92	0.96	0.55	0.11	11.91
Mg	12	0.80	0.83	1.02	0.12	14.82
Fe	26	0.63	0.66	0.35	0.10	15.95
Ni	28	0.34	0.36	0.18	0.09	26.53
Ti	22	0.00	0.00	0.00	0.00	0.88
Nb	41	0.00	0.00	0.00	0.00	1.59
		96.48	**100.00**	**100.00**		

(c)

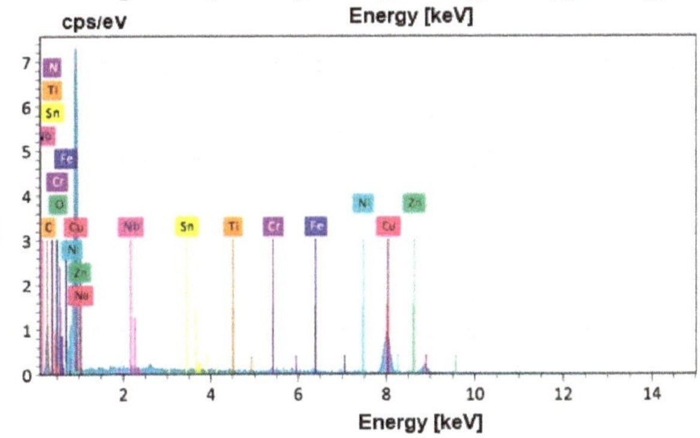

Element	At. No.	Mass [%]	Mass Norm. [%]	Atom [%]	abs. error [%] (1 sigma)	rel. error [%] (1 sigma)
Cu	29	72.87	69.61	34.29	2.93	4.01
C	6	19.44	18.57	48.39	4.71	24.23
O	8	7.80	7.45	14.58	1.95	25.03
Zn	30	2.19	2.09	1.00	0.35	16.08
Fe	26	0.68	0.65	0.36	0.12	17.15
Cr	24	0.56	0.53	0.32	0.10	17.77
Ni	28	0.52	0.50	0.27	0.13	24.32
N	7	0.28	0.27	0.60	0.40	143.17
Ti	22	0.27	0.26	0.17	0.07	26.55
Nb	41	0.04	0.04	0.01	0.02	42.43
Sn	50	0.03	0.02	0.01	0.01	55.17
Na	11	0.00	0.00	0.00	0.00	4.18
		104.69	**100.00**	**100.00**		

Figure 13. EDS Image of Machined Tool Electrodes (**a**) Uncoated under trial 5 (**b**) Uncoated trial 3 (**c**) Uncoated trial 8.

In Figures 11 and 12, the EDS results on the workpiece from ECMM both uncoated and Ag-coated electrodes under two different experiments are shown. More specifically, the conditions of trial 5 were less harsh than the ones of trial 8, with the latter resulting in higher MRR and overcut both in the case with the uncoated and the coated electrode.

Regarding the EDS results for the workpiece after trial 5, when the Ag-coated electrode was used, it can be seen from Figure 11, that in comparison to the initial composition of the workpiece, the percentages of various elements have changed, with Ni, Fe, and Mo being reduced while Cu and C were increased. Whereas the lower percentages of Ni, Fe, and Mo can be attributed to the removed material and the slightly increased percentage of Cu can be attributed to the electrode material, the 4.70% of residual C on the surface of micro-holes shows that the presence of carbon content in electrolyte dissolution has reflected its effect on the workpiece during the ECM machining process, leading to less material removal from the workpiece using the better combination of the low input parameters and the silver-coated tool electrode. In the case of the uncoated electrode, most results are similar, except for the larger amount of C on the workpiece, which was released from the dissolution of the electrolyte.

(a)

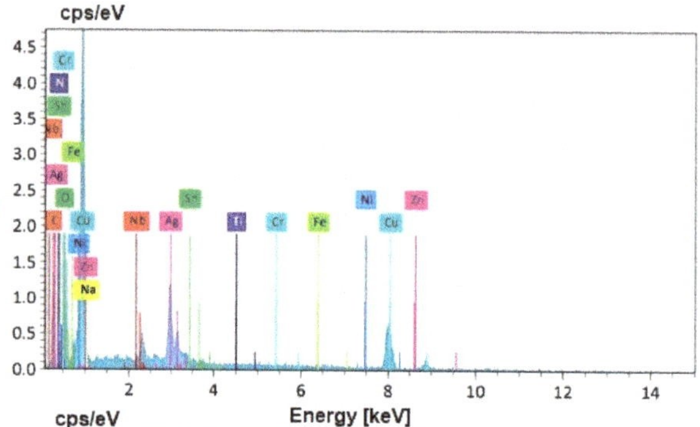

Element	At. No.	Mass [%]	Mass Norm. [%]	Atom [%]	abs. error [%] (1 sigma)	rel. error [%] (1 sigma)
Cu	29	50.61	50.92	25.90	2.09	4.12
Ag	47	18.72	18.83	5.64	0.72	3.84
C	6	18.21	18.32	49.29	3.84	21.11
O	8	8.38	8.43	17.03	2.07	24.74
Zn	30	1.88	1.89	0.94	0.29	15.44
Nb	41	0.86	0.86	0.30	0.10	12.23
Na	11	0.27	0.27	0.38	0.08	29.37
N	7	0.17	0.17	0.39	0.27	162.26
Cr	24	0.17	0.17	0.10	0.06	37.52
Sn	50	0.13	0.13	0.04	0.05	42.77
Fe	26	0.00	0.00	0.00	0.00	0.92
Ni	28	0.00	0.00	0.00	0.00	1.04
Ti	22	0.00	0.00	0.00	0.00	0.88
		99.38	**100.00**	**100.00**		

(b)

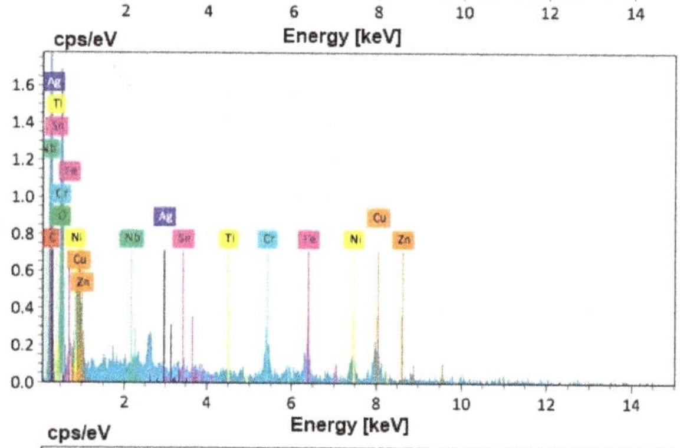

Element	At. No.	Mass [%]	Mass Norm. [%]	Atom [%]	abs. error [%] (1 sigma)	rel. error [%] (1 sigma)
C	6	36.12	34.94	57.12	7.79	21.57
O	8	24.71	23.90	29.34	5.41	21.88
Cu	29	21.75	21.04	6.50	1.47	6.77
Ni	28	7.38	7.14	2.39	0.66	8.91
Fe	26	6.24	6.04	2.12	0.48	7.75
Cr	24	5.39	5.21	1.97	0.38	6.99
Zn	30	1.51	1.46	0.44	0.39	25.77
Ti	22	0.30	0.29	0.12	0.09	29.16
Nb	41	0.00	0.00	0.00	0.00	1.59
Sn	50	0.00	0.00	0.00	0.00	0.98
Ag	47	0.00	0.00	0.00	0.00	1.12
		103.39	**100.00**	**100.00**		

(c)

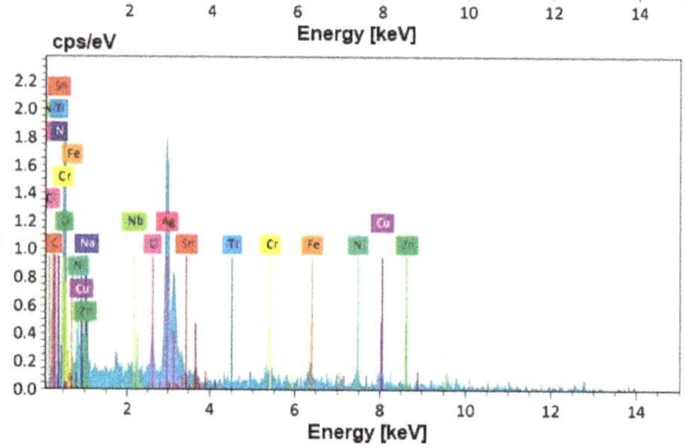

Element	At. No.	Mass [%]	Mass Norm. [%]	Atom [%]	abs. error [%] (1 sigma)	rel. error [%] (1 sigma)
Ag	47	50.31	39.73	10.53	2.02	4.02
O	8	36.46	28.79	51.47	8.62	23.64
C	6	13.12	10.36	24.67	4.02	30.65
Cu	29	7.14	5.64	2.54	0.83	11.57
Na	11	4.51	3.56	4.43	0.50	11.15
Ni	28	4.31	3.40	1.66	0.55	12.75
Fe	26	3.91	3.08	1.58	0.45	11.46
Cr	24	3.12	2.47	1.36	0.35	11.17
Cl	17	2.16	1.70	1.37	0.22	10.24
Sn	50	1.06	0.84	0.20	0.21	19.60
Zn	30	0.54	0.43	0.19	0.25	46.43
Nb	41	0.00	0.00	0.00	0.00	1.59
Ti	22	0.00	0.00	0.00	0.00	0.88
N	7	0.00	0.00	0.00	0.00	10.00
		126.63	**100.00**	**100.00**		

Figure 14. EDS Image of Machined Tool Electrodes (**a**) Ag-coated under trial 5 (**b**) Ag-coated trial 3 (**c**) Ag-coated trial 8.

On the other hand, ECMM under harsher conditions, such as the ones of trial 8, shows more considerable alteration of the composition of the workpiece surface. From Figure 12, it can be seen that the percentages of Ni and Fe are lower than the ones of trial 5 for the same type of electrode, while the percentage of C on the workpiece is higher in the case of the coated electrode and slightly lower in the case of uncoated electrode.

From EDS analysis of the tool electrode as shown in Figures 13 and 14, the chemical composition of the copper and Ag-coated copper electrodes was similar to the initial one, and no elements relevant to the workpiece elements composition (Ni, Fe, Cr) were observed under the medium parametric condition of trial 5 (12 V, 25 g/L, 70 Hz, and duty factor of 33%). Due to the lower percentage of carbon and oxygen in the alloy, the reaction of silver (coating) with carbonic oxide available in the electrolyte was reduced. However, the lower duty cycle means that the energy will be applied to the workpiece in a shorter time than in the case of trial 8, leading to higher removal of Ag-coating. Although in slightly more intense conditions, such as the ones of trial 8, the reaction of Silver (coating) with the carbonic oxide in the electrolyte is higher, the higher duty cycle finally leads to a lower quantity of Ag coating being removed. Due to over machining of the tool electrode in case 8, after the removal of silver coating from the tool surface, elements from the workpiece (Ni, Cr, Fe) were embedded in the tool electrode surface in larger quantities than in trial 8, namely 4.31%, 3.12%, and 3.91% instead of 0%, 0.17%, and 0%, respectively.

4. Conclusions

In this investigation, micro-holes were created on Incoloy 825 alloy by ECMM process with copper and Silver-coated copper tool electrodes under various combinations of voltage, electrolyte concentration, pulse frequency, and duty factor in order to achieve the most desirable performance parameters regarding MRR, overcut and conicity of the produced holes. The following results were framed from experiments conducted with uncoated and silver-coated copper tool electrodes.

1. The MRR obtained using the Silver-coated copper tool was higher than the uncoated tool electrode in most cases, because of the high electrical conductivity obtained in the Silver-coated copper tool electrode. For both the actual and theoretical MRR, the pulse frequency was the most important parameter when the uncoated electrode was used, whereas the electrolyte concentration was the most important parameter, when the coated electrode was used. From these results, it can be recommended that, in order to achieve higher productivity in the ECMM process, a coated electrode and high values of voltage, electrolyte concentration, and duty cycle should be used, along with relatively low-frequency values.
2. The overcut results obtained using the silver-coated copper tool electrode were better when compared to the uncoated tool electrode, since the Silver-coated copper tool electrode has a fine grain structure that restricts the flow of electrons. For the overcut, the most important parameter was the duty factor when the uncoated electrode was used and the voltage when the coated electrode was used. From these results, it can be recommended that, in order to achieve lower overcut, a coated electrode and low values of voltage, electrolyte concentration, and duty cycle should be used, along with high-frequency values.
3. The conicity levels are generally higher with an uncoated copper tool electrode. For conicity, when the uncoated electrode was used the duty factor was the most important parameter, whereas when the coated electrode was used, the voltage was the most important parameter. In order to minimize conicity, it is recommended that a coated electrode should be used at high voltage, moderate electrolyte concentration, low frequency, and low-duty cycle conditions.
4. The silver-coated copper tool electrode tends to have fewer cracks, pit formations, and corroded regions than the uncoated copper tool electrode because of silver coating, which acted as a corrosion shield when coming into contact with carbon contented electrolyte dissolution.
5. The EDS analysis revealed that when a silver-coated copper tool electrode was used, it could produce fewer particles of tool electrode over the machined surface without affecting real constituents of the workpiece specimens. Under harsher conditions, the embedment of workpiece elements on the tool electrode is increased. Finally, the removal of Ag coating was correlated to the energy input, mainly represented by the duty cycle, as higher amount of Ag was removed at lower duty cycle values, when the energy is applied in a shorter duration.

Author Contributions: Conceptualization, G.T. and M.T.; methodology, G.T., M.T., P.I.A. and M.J.; software, M.T. and E.L.P.; validation, N.E.K., E.L.P. and P.K.-O.; formal analysis, G.T., M.T., N.E.K., E.L.P. and P.K.-O.; investigation, G.T., M.T., P.I.A. and M.J.; resources, M.T.; data curation, G.T. and M.T.; writing—original draft preparation, G.T., M.T., P.I.A. and M.J.; writing—review and editing, N.E.K., E.L.P. and P.K.-O.; visualization, G.T., M.T. and E.L.P.; supervision, M.T.; project administration, M.T.; funding acquisition, N.E.K. All authors have read and agreed to the published version of the manuscript.

Funding: This research received no external funding.

Institutional Review Board Statement: Not applicable.

Informed Consent Statement: Not applicable.

Data Availability Statement: The data presented in this study are available on request from the corresponding author.

Conflicts of Interest: The authors declare no conflict of interest.

References

1. Kangazian, J.; Sayyar, N.; Shamanian, M. Influence of Microstructural Features on the Mechanical Behavior of Incoloy 825 Welds. *Metallogr. Microstruct. Anal.* **2017**, *6*, 190–199. [CrossRef]
2. Ding, Q.; Lao, Z.; Wei, H.; Li, J.; Bei, H.; Zhang, Z. Site Occupancy of Alloying Elements in Γ′ Phase of Nickel-Base Single Crystal Superalloys. *Intermetallics* **2020**, *121*, 106772. [CrossRef]
3. Wei, C.-N.; Bor, H.-Y.; Chang, L. The Effects of Carbon Content on the Microstructure and Elevated Temperature Tensile Strength of a Nickel-Base Superalloy. *Mater. Sci. Eng. A* **2010**, *527*, 3741–3747. [CrossRef]
4. Choudhury, B.; Chandrasekaran, M.; Devarasiddappa, D. Development of ANN Modelling for Estimation of Weld Strength and Integrated Optimization for GTAW of Inconel 825 Sheets Used in Aero Engine Components. *J. Braz. Soc. Mech. Sci. Eng.* **2020**, *42*, 308. [CrossRef]
5. Reed, R.C. *The Superalloys, Fundamentals and Applications*; Cambridge University Press: Cambridge, UK, 2006.
6. Sims, C.; Stoloff, N.S.; Hagel, W.C. (Eds.) *Superalloys II*; John Wiley & Sons: Hoboken, NJ, USA, 1987.
7. Geethapriyan, T.; Muthuramalingam, T.; Moiduddin, K.; Mian, S.M.; Alkhalefah, H.; Umer, U. Performance Analysis of Electrochemical Micro Machining of Titanium (Ti-6Al-4V) Alloy under Different Electrolytes Concentrations. *Metals* **2021**, *11*, 247. [CrossRef]
8. Geethapriyan, T.; Muthuramalingam, T.; Kalaichelvan, K. Influence of Process Parameters on Machinability of Inconel 718 by Electrochemical Micromachining Process using TOPSIS Technique. *Arab. J. Sci. Eng.* **2019**, *44*, 7945–7955. [CrossRef]
9. Rajurkar, K.P.; Sundaram, M.M.; Malshe, A.P. Review of Electrochemical and Electrodischarge Machining. *Procedia CIRP* **2013**, *6*, 13–26. [CrossRef]
10. Geethapriyan, T.; Muthuramalingam, T.; Moiduddin, K.; Alkhalefah, H.; Mahalingam, S.; Karmiris-Obratański, P. Multiobjective Optimization of Heat-Treated Copper Tool Electrode on EMM Process Using Artificial Bee Colony (ABC) Algorithm. *Materials* **2022**, *15*, 4831. [CrossRef]
11. Liu, S.; Geethapriyan, T.; Muthuramalingam, T.; Shanmugam, R.; Ramoni, M. Influence of heat-treated Cu-Be electrode on machining accuracy in ECMM with Monel 400 alloy. *Arch. Civil. Mech. Eng.* **2022**, *22*, 154. [CrossRef]
12. Li, H.; Gao, C.; Wang, G.; Qu, N.; Zhu, D. A Study of Electrochemical Machining of Ti-6Al-4V in $NaNO_3$ Solution. *Sci. Rep.* **2016**, *6*, 35013. [CrossRef]
13. Harugade, M.; Kavade, M.V.; Hargude, N. Effect of Electrolyte Solution on Material Removal Rate in Electrochemical Discharge Machining. *IOSR J. Mech. Civ. Eng.* **2013**, *5*, 1–8.
14. Manikandan, N.; Kumanan, S.; Sathiyanarayanan, C. Multiple Performance Optimization of Electrochemical Drilling of Inconel 625 Using Taguchi Based Grey Relational Analysis. *Eng. Sci. Technol. Int. J.* **2017**, *20*, 662–671. [CrossRef]
15. Krishnan, R.; Duraisamy, S.; Palanisamy, P.; Veeramani, A. Optimization of the Machining Parameters in the Electrochemical Micro-Machining of Nickel. *Mater. Tehnol.* **2018**, *52*, 253–258. [CrossRef]
16. Geethapriyan, T.; Kalaichelvan, K.; Muthuramalingam, T.; Rajadurai, A. Performance analysis of process parameters on machining α-β titanium alloy in electrochemical micromachining process. *Proc. Inst. Mech. Eng. B J. Eng. Manuf.* **2018**, *232*, 1577–1589. [CrossRef]
17. Liu, G.; Li, Y.; Kong, Q.; Tong, H.; Zhong, H. Silicon-Based Tool Electrodes for Micro Electrochemical Machining. *Precis. Eng.* **2018**, *52*, 425–433. [CrossRef]
18. Shanmugam, R.; Ramoni, M.; Thangamani, G.; Thangaraj, M. Influence of Additive Manufactured Stainless Steel Tool Electrode on Machinability of Beta Titanium Alloy. *Metals* **2021**, *11*, 778. [CrossRef]
19. Egashira, K.; Hayashi, A.; Hirai, Y.; Yamaguchi, K.; Ota, M. Drilling of Microholes Using Electrochemical Machining. *Precis. Eng.* **2018**, *54*, 338–343. [CrossRef]

20. Mithu, M.A.H.; Fantoni, G.; Ciampi, J.; Santochi, M. On How Tool Geometry, Applied Frequency and Machining Parameters Influence Electrochemical Microdrilling. *CIRP J. Manuf. Sci. Technol.* **2012**, *5*, 202–213. [CrossRef]
21. Swain, A.K.; Sundaram, M.M.; Rajurkar, K.P. Use of Coated Microtools in Advanced Manufacturing: An Exploratory Study in Electrochemical Machining (ECM) Context. *J. Manuf. Process.* **2012**, *14*, 150–159. [CrossRef]
22. Bhattacharyya, B.; Malapati, M.; Munda, J.; Sarkar, A. Influence of Tool Vibration on Machining Performance in Electrochemical Micro-Machining of Copper. *Int. J. Mach. Tools Manuf.* **2007**, *47*, 335–342. [CrossRef]
23. Ghoshal, B.; Bhattacharyya, B. Generation of Microfeatures on Stainless Steel by Electrochemical Micromachining. *Int. J. Adv. Manuf. Technol.* **2015**, *76*, 39–50. [CrossRef]
24. Wielage, B.; Mäder, T.; Fürderer, B.; Nestler, D.; Reinecke, H. Carbon Fibres Used as Electrode Tools for Micro-Electrochemical Machining. In Proceedings of the 17th International Conference on Composite Materials, Edinburgh, UK, 27–31 July 2009.
25. Chen, X.; Xu, Z.; Zhu, D.; Fang, Z.; Zhu, D. Experimental Research on Electrochemical Machining of Titanium Alloy Ti60 for a Blisk. *Chin. J. Aeronaut.* **2016**, *29*, 274–282. [CrossRef]
26. Thanigaivelan, R.; Arunachalam, R. Study on Influence of Tool Design on Electrochemical Micromachining. *Manuf. Technol. Today* **2011**, *10*, 3–10.
27. Anasane, S.S.; Bhattacharyya, B. Experimental Investigation into Micromilling of Microgrooves on Titanium by Electrochemical Micromachining. *J. Manuf. Process.* **2017**, *28*, 285–294. [CrossRef]
28. Duraisamy, S.; Arularasu, M.; Ganesan, K. A Study on Electrochemical Micromachining of Super Duplex Stainless Steel for Biomedical Filters. *ARPN J. Eng. Appl. Sci.* **2012**, *7*, 517–523.
29. Ayyappan, S.; Sivakumar, K. Experimental Investigation on the Performance Improvement of Electrochemical Machining Process Using Oxygen-Enriched Electrolyte. *Int. J. Adv. Manuf. Technol.* **2014**, *75*, 479–487. [CrossRef]
30. Soni, S.K.; Thomas, B. A Comparative Study of Electrochemical Machining Process Parameters by Using GA and Taguchi Method. *IOP Conf. Ser. Mater. Sci. Eng.* **2017**, *263*, 062038. [CrossRef]
31. Uttarwar, S.S.; Chopde, I.K. A Study of Influence of Electrochemical Process Parameters on the Material Removal Rate and Surface Roughness of SS AISI 304. *Int. J. Comput. Eng. Res.* **2013**, *3*, 189–197.
32. Geethapriyan, T.; Kalaichelvan, K. Experimental Investigation of Electrochemical Micromachining Process Parameters on Pure-Titanium Using Taguchi-Grey Relational Analysis. *Appl. Mech. Mater.* **2016**, *852*, 198–204. [CrossRef]
33. Geethapriyan, T.; Kalaichelvan, K.; Muthuramalingam, T. Influence of Coated Tool Electrode on Drilling Inconel Alloy 718 in Electrochemical Micro Machining. *Procedia CIRP* **2016**, *46*, 127–130. [CrossRef]
34. Muthuramalingam, T.; Thangamani, G.; Kalaichelvan, K. Multi Performance Optimization of Electrochemical Micro-Machining Process Surface Related Parameters on Machining Inconel 718 Using Taguchi-Grey Relational Analysis. *Metall. Ital.* **2016**, *2016*, 13–19.
35. Nugroho, A.W.; Sudarisman; Nurahman, M.B.; Septiaji, P. Overcut and material removal rate on electrochemical machining of aluminum and stainless steel using isolated brass electrode. In Proceedings of the 2nd International Conference of Industrial, Mechanical, Electrical, Chemical Engineering (ICIMECE), Yogyakarta, Indonesia, 6–7 October 2016.
36. Das, A.; Padhan, S.; Das, S.R. Analysis on hole overcut during micro-EDM of Inconel 718. *Mater. Today Proceed.* **2022**, *56*, 29–35. [CrossRef]
37. Sathish, T. Experimental investigation of machined hole and optimization of machining parameters using electrochemical machining. *J. Mater. Res. Technol.* **2019**, *8*, 4354–4363. [CrossRef]
38. Tajdari, M.; Chavoshi, S.Z. Prediction and analysis of radial overcut in holes drilled by electrochemical machining process. *Cent. Eur. J. Eng.* **2013**, *3*, 466–474. [CrossRef]
39. Aravind, S.; Hiremath, S.S. Machining of holes on SS316L with solid and hollow tool electrodes. *Mater. Manuf. Process.* **2022**, *37*, 1859–1870. [CrossRef]
40. Deepak, J.; Hariharan, P. Investigation of electrochemical machining on SS304 using NaCl and $NaNO_3$ as electrolyte. *Mater. Manuf. Process.* **2022**, *37*, 1790–1803.
41. Teimouri, R.; Baseri, H. Study of tool wear and overcut in EDM process with rotary tool and magnetic field. *Adv. Tribiol.* **2012**, *2012*, 895918. [CrossRef]
42. Machno, M.; Matras, A.; Szkoda, M. Modelling and analysis of the effect of EDM-Drilling parameters on the machining performance of Inconel 718 using the RSM and ANNs methods. *Materials* **2022**, *15*, 1152. [CrossRef]
43. D'Urso, G.; Merla, C. Workpiece and electrode influence on micro-EDM drilling performance. *Precis. Eng.* **2014**, *38*, 903–914. [CrossRef]
44. Wen, Z.; Pei, H.; Zhang, C.; Wang, B. Analysis of surface quality of multi-film cooling holes in nickel-based single crystal superalloy. *Mater. Sci. Technol.* **2016**, *32*, 1845–1854. [CrossRef]
45. Vivekkumar, P.; Soundrapandian, E.; Tajdeen, A.; Prashanth, T. Experimental study of Electrochemical micromachining on Titanium (Ti-6Al-4V) Alloy. In *Advances in Materials Research. Springer Proceedings in Materials;* Kumaresan, G., Shanmugam, N.S., Dhinakaran, V., Eds.; Springer Nature: Singapore, 2021; pp. 301–308.
46. Sen, M.; Shan, H.S. A review of electrochemical macro- to micro-hole drilling processes. *Int. J. Mach. Tools Manuf.* **2005**, *45*, 137–152. [CrossRef]
47. Machno, M.; Bogucki, R.; Szkoda, M.; Bizoń, W. Impact of the deionized water on making high aspect ratio holes in the Inconel 718 Alloy with the use of Electrical Discharge Drilling. *Materials* **2020**, *13*, 1476. [CrossRef] [PubMed]

48. Phan, N.H.; Muthuramalingam, T.; Nguyen, T.L.; Tran, V.D. Effect of ultrasonic low frequency vibration and its direction on machinability in WEDM process. *Mater. Manuf. Process.* **2022**, *37*, 1045–1051. [CrossRef]
49. Muthuramalingam, T.; Moiduddin, K.; Akash, R.; Krishnan, S.; Mian, S.H.; Ameen, W.; Alkhalefah, H. Influence of process parameters on dimensional accuracy of machined Titanium (Ti-6Al-4V) alloy in Laser Beam Machining Process. *Opt. Laser. Technol.* **2020**, *132*, 106494. [CrossRef]

coatings

Article

ZIF-67 Derived Co-Ni/C Composites for High-Efficiency Broadband Microwave Absorption

Qingwei Liu [1], Yiming Zhao [1,*], Zhao Fan [1], Jiali Yu [1] and Min Zeng [2]

[1] State-Owned Wuhu Machinery Factory, Wuhu 241000, China
[2] School of Materials Science and Engineering, Beihang University, Beijing 100191, China
* Correspondence: zhaoyiming@mail.nwpu.edu.cn

Abstract: The development of broadband and high-efficiency broadband microwave-absorbing materials is one of the keys to resisting electromagnetic radiation and electromagnetic interference. In this paper, we used zeolitic imidazolate framework-67 (ZIF-67) as the template to obtain Ni-doped ZIF-67 via $Ni(NO_3)_2$ etching followed by pyrolysis to prepare Co-Ni/C composites successfully. The morphology and microstructure of Co-Ni/C composites can be well-tuned by altering the Ni doping content. ZIF-67-derived Co-Ni/C porous materials exhibit extremely strong microwave absorption performance thanks to dielectric loss, magnetic loss, and the synergistic effect between different components. When the dosage of $Ni(NO_3)_2$ used reaches 0.32 g, the obtained composite possesses the optimal absorbing properties with a maximum reflection loss (*RL*) of −72.88 dB and an effective bandwidth (f_e, representing $RL \leq -10$ dB) of 7.04 GHz (9.76–16.8 GHz) corresponding to a thickness of 2.53 mm. The results of this work indicate that Co-Ni/C composites have potential application value in the field of microwave absorption.

Keywords: microwave absorption; ZIF-67; Ni doping; composite; electromagnetic parameters

1. Introduction

In recent decades, microwave-absorbing materials have attracted increasing interest, for they can convert electromagnetic energy into heat energy, effectively protecting human health, the safety of military facilities, communication signals, etc. [1]. There are many kinds of microwave-absorbing materials, including ferrite materials, metal-based materials, carbon-based materials, ceramic materials, etc. [2]. However, single-component microwave absorbers usually have only a single microwave loss mechanism of magnetic loss (such as magnetic metal) or dielectric loss (such as carbon-based material), which means it is hard for them to satisfy the requirements of microwave absorption. In recent years, metal/carbon composites have aroused extensive research interest as lightweight microwave absorbers because they benefit from the properties of both carbon materials and metal materials, generating magnetic loss and dielectric loss simultaneously.

As a type of functional material, metal–organic frameworks (MOFs) have been utilized in many fields such as adsorption, catalysis, and sensing due to their diverse structures, high specific surface area, high porosity, and ease of fabrication [3]. Meanwhile, the microstructure of MOFs is considered to be an ideal model for microwave-absorbing materials with the advantages of light weight, excellent absorption performance, and ease of fabrication [4,5]. Therefore, the design and preparation of MOF derivatives in the field of microwave absorption is a hot topic.

Zeolitic imidazolate frameworks (ZIFs), as a class of MOFs, are composed of transition-metal cations (M) and imidazole-based ligands (lm). Among them, ZIF-67 has received a lot of attention in the microwave absorption field [6]. Zheng's group first used ZIF-67 as a precursor to obtain Co/C composite absorbers by high-temperature pyrolysis [7]. They studied the effect of pyrolysis temperature on microwave absorption performance, indicating the microwave absorber prepared at 500 °C has optimal performance, with a maximum *RL* of −35.3 dB and a f_e of 5.80 GHz (8.40 GHz–14.20 GHz). Shortly afterward, microwave-absorbing materials obtained by high-temperature pyrolysis using MOFs as precursors aroused great research interest [8–10]. Che et al. constructed shelled ZIF-67 rhombic dodecahedral cages from tannic acid followed by pyrolysis to obtain hollow Co@N-doped carbon nanocages with uniform heterojunction and hierarchical micro-pores. The maximum reflection loss at 2.4 mm is −60.6 dB, and the corresponding f_e at 1.9 mm reaches 5.1 GHz with 10 wt% filler [11]. Yan et al. successfully prepared CoS_2/ N-doped carbon nanotube composites (CoS_2/NCNT) derived from MOF, indicating CoS_2/ NCNT with 50% loadings exhibits optimal microwave absorption performance, the f_e almost covering the entire X-band with a thickness of only 1.6 mm [12]. However, further improving the microwave absorption performance of ZIF-67-derived microwave-absorbing materials, especially the design and preparation of broadband (>6 GHz) materials, still needs to be further explored.

In recent years, hollow nanostructures have attracted extensive attention owing to their strong microwave absorption properties [13]. On one hand, hollow nanostructures have a large specific surface area, which is conducive to enhancing the impedance matching performance as well as facilitating multiple scattering of electromagnetic waves. On the other hand, hollow nanostructures introduce more defects and polarization centers, thereby increasing the dielectric loss.

Herein, Ni-doped ZIF-67 was prepared by etching nano-cubic ZIF-67 with $Ni(NO_3)_2$ followed by high-temperature in-situ pyrolysis in Ar/H_2 atmosphere to obtain light-weight and strongly absorbing Co-Ni/C composites as microwave absorbers. Interestingly, the Co-Ni/C composite exhibits extremely strong microwave-absorption properties, with a maximum *RL* of −72.88 dB and the corresponding f_e reaching 7.04 GHz, indicating it can be used as an excellent microwave absorber in stealth structure design.

2. Experiment and Characterization

2.1. Synthesis of ZIF-67

$Co(NO_3)_2{\cdot}6H_2O$ (58 mg) and 1 mg cetyltrimethylammonium bromide (CTAB) were added into 2 mL of aqueous solution to form a uniform solution. Then, the mixture was injected into 14 mL of aqueous solution with 908 mg of 2-methylimidazole and vigorously stirred for 20 min. Then, the precipitate yielded was collected by thoroughly washing with absolute alcohol several times before vacuum drying at room temperature overnight.

2.2. Synthesis of Ni-Doped ZIF-67

The as-prepared ZIF-67 nanocubes (0.16 g) were first dispersed in 100 mL of absolute alcohol and then $Ni(NO_3)_2{\cdot}6H_2O$ with different amounts (0 g, 0.16 g, 0.32 g, 0.64 g) were added into the suspension. After stirring for 20 min, Ni-doped ZIF-67 layered double hydroxides (LDHs) were formed, denoted as Pre-0, Pre-16, Pre-32, and Pre-64, respectively. Then, the products were centrifuged with absolute alcohol several times.

2.3. Synthesis of Co-Ni/C

To obtain Co-Ni/C materials, the as-prepared Ni-doped ZIF-67 was heated at 800 °C for 2 h in an Ar/H_2 (95%/5% in volume) atmosphere. The series of products were named S-0, S-16, S-32, and S-64, annealed from Pre-0, Pre-16, Pre-32, and Pre-64, respectively. The whole synthesis process is illustrated in Figure 1.

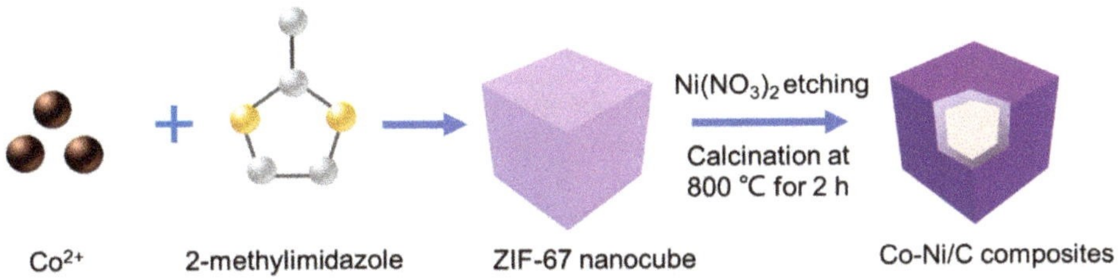

Figure 1. Schematic illustration of the fabrication of Co-Ni/C composites.

2.4. Characterization

The relative complex permittivity and permeability were determined by a vector network analyzer (VNA, N5234A, Agilent, Santa Clara, CA, USA) in the frequency range of 2–18 GHz using a rectangular waveguide. The samples for the electromagnetic parameters test were prepared as follows: Co-Ni/C composites were mixed with epoxy resin at a mass ratio of 3:7 and pressed into a rectangle structure for the electromagnetic parameters test. As lightweight materials, MOF derivatives only need very little dosage to achieve high-performance microwave absorption, and excessive addition will also lead to poor performance, so the filling amount of the filler was chosen to be 30%. The surface morphologies of the samples were observed by scanning electron microscopy (SEM, JSM7500, JEOL, Akishima, Japan).

3. Results and Discussion

3.1. Microstructure of Ni-Doped ZIF-67

Figure 2 exhibits the SEM of ZIF-67 precursor treated with different amounts of $Ni(NO_3)_2$. The synthesized ZIF-67 precursor without $Ni(NO_3)_2$ treatment in Figure 2a has a regular cubic shape with a size of approximately 500 nm. After etching with different amounts of $Ni(NO_3)_2$, a variety of microstructures were obtained. As shown in Figure 2b, when the amount of $Ni(NO_3)_2$ is 0.16 g, the nanocubes basically maintain the original morphology with slight surface roughness and shrinkage. As the amount of $Ni(NO_3)_2$ increases, broken shell cubes are found in Figure 2c, indicating the interior of ZIF-67 has been etched and then turned into hollow cubes, which increase the surface roughness. Surface roughness plays a crucial role in microwave absorption properties since a high surface area provides more active sites for electromagnetic wave reflection and scattering and promotes the repeated absorption process of electromagnetic waves. In addition, a rough surface is more suitable for impedance matching. When the amount of $Ni(NO_3)_2$ increased to 0.64 g, the nanocubes were broken, and it was difficult to maintain the original morphology in Figure 2d. The phenomenon mentioned above can be explained as follows: ZIF-67 nanocubes are reacted with $Ni(NO_3)_2$ to form Ni-Co LDH on the surface of ZIF-67 nanocubes, while the inner part of ZIF-67 diffuses out of the nanocubes to form ZIF-67@LDH nanoboxes [14].

3.2. Microstructure and Microwave Absorption Properties of Co-Ni/C

Figure 3 exhibits the SEM of Ni-doped ZIF-67 after 800 °C pyrolysis. The surface of the S-0 sample is covered with coiled carbon nanotubes, which is in accord with the research results of Yin et al. [15], indicating the catalytic effect of Co promotes the transformation of carbon in ZIF-67 to carbon nanotubes. When etched by $Ni(NO_3)_2$, the carbon source is correspondingly reduced, and only a few short carbon nanotubes are observed as circled in red in Figure 3b,c. When the content of $Ni(NO_3)_2$ is further increased, almost no carbon nanotubes are observed, followed by a significant agglomeration in Figure 3d. In general, particle aggregation leads to a decrease in the surface area of the materials and a poor impedance match with the air, which is not conducive to the absorption of electromagnetic waves.

Figure 2. SEM of ZIF-67 precursor treated with different amounts of $Ni(NO_3)_2$: (**a**) Pre-0, (**b**) Pre-16, (**c**) Pre-32, (**d**) Pre-64.

Transmission line theory and metal backplane model were used to calculate *RL* and corresponding impedance of the test samples, which can be expressed as [16]:

$$RL = 20\log_{10}\left|\frac{Z_1 - Z_0}{Z_1 + Z_0}\right| \quad (1)$$

$$Z_1 = \sqrt{\mu_1/\varepsilon_1}\tanh\left[j2\pi f d_1\sqrt{\mu_1\varepsilon_1}/c\right] \quad (2)$$

where Z_0 is the impedance of free space, Z_{in} is the input impedance of the absorber, and ε_1 and μ_1 are the relative complex permittivity and permeability of the sample, respectively. Additionally, d_1 is the thickness of the sample, f is the frequency of the microwave, and c represents the speed of light in free space.

Figure 4 shows the calculated results of the frequency-dependent *RL* properties of Co-Ni/C composites with variable absorber thicknesses (d) from 1.0 to 5.0 mm. Although all the samples can work as microwave absorbers, their properties are greatly distinguishable. The maximum *RL* of sample S-0 is −58.85 dB (1.84 mm, 13.12 GHz) while the corresponding effective absorption bandwidth is only 3.84 GHz (1.68 mm, 12.72–16.56 GHz), indicating that although strong absorption of microwaves is achieved, the effective absorption bandwidth is relatively narrow. With the increase in the amount of $Ni(NO_3)_2$, the *RL* value of samples can be significantly enhanced, and the maximum *RL* for S-16, S-32, and S-64 reach up to −72.88 dB (2.7 mm, 12.24 GHz), −68.12 dB (2.21 mm, 14.24 GHz), and −59.05 dB (1.47 mm, 17.28 GHz), with the corresponding f_e reaching 6.08 GHz (2.34 mm, 11.92–18 GHz), 7.04 GHz (2.53 mm, 9.76–16.8 GHz), and 3.68 GHz (1.49 mm, 12.4–16.08 GHz), respectively. The results indicate that tunable microwave absorption performance can be achieved by changing the amount of $Ni(NO_3)_2$.

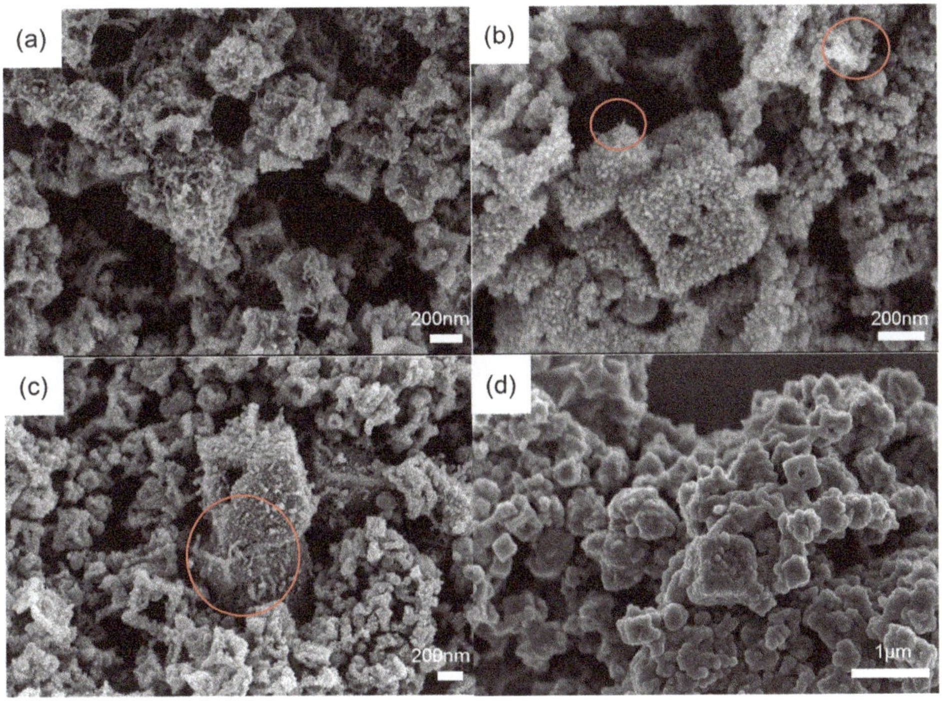

Figure 3. SEM of Ni-doped ZIF-67 after 800 °C pyrolysis: (**a**) S-0, (**b**) S-16, (**c**) S-32, (**d**) S-64.

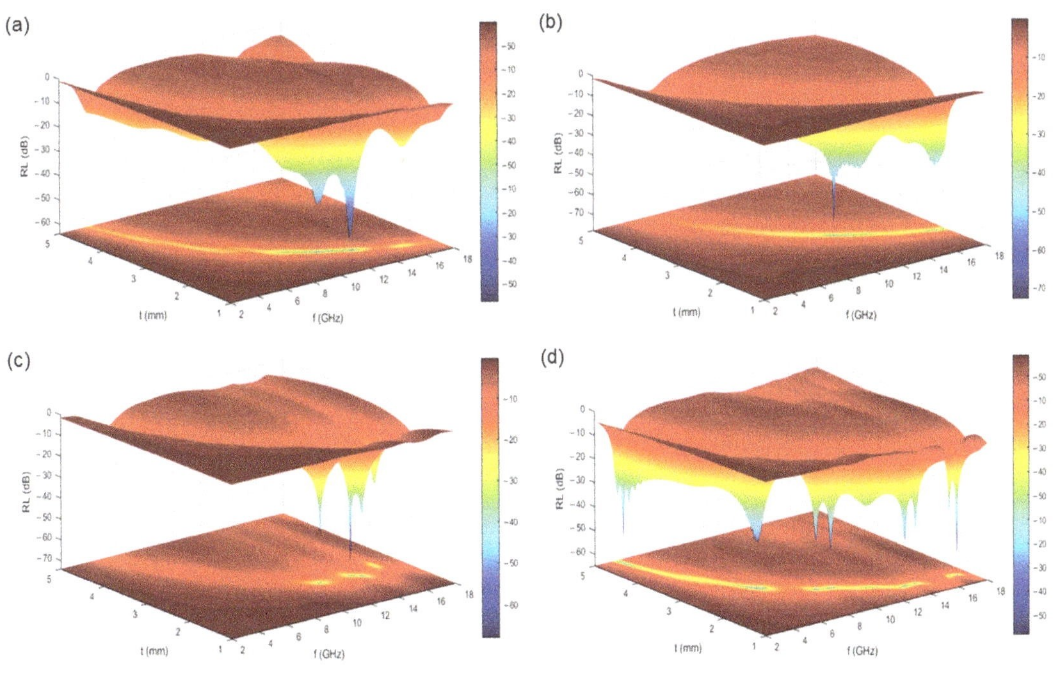

Figure 4. *RL* of Ni-doped ZIF-67 after 800 °C pyrolysis: (**a**) S-0, (**b**) S-16, (**c**) S-32, (**d**) S-64.

3.3. Microwave Absorption Mechanism of Co-Ni/C

RL is determined by electromagnetic parameters, which are the key to understanding the electromagnetic properties of materials. Thus, analyzing the relationship between electromagnetic parameters and frequency is helpful for insight into the underlying mechanism of microwave absorption. In the frequency range of 2–18 GHz, the electromagnetic parameters of Co-Ni/C are measured, and the corresponding results are exhibited in Figure 5. The real part (ϵ' and μ') represents the ability to store microwave energy, and the corresponding imaginary part (ϵ'' and μ'') represents the ability to consume microwave energy [17]. In Figure 5a, ε' decreases with increasing frequency which can be due to the increased hysteresis effect of the polarized charge. The S-0 and S-64 samples have larger ε' values, indicating the influence of material composition is crucial. According to the free electron theory, the S-0 sample has a high degree of graphitization under the catalysis of Ni, corresponding to a high ϵ' value [18]. Generally, the dielectric loss of material in the GHz frequency band is attributed to interface polarization, conductivity loss, and dipole polarization [19]. In alternating electromagnetic fields, space charges will accumulate at the heterogeneous interface between different mediums, resulting in interface polarization. The interfaces in Co-Ni/C include the Co-Ni interface, Co-C interface, and Ni-C interface as well as the interface between Co-Ni/C and epoxy resin. The presence of C and N defects in the sample results in an asymmetric distribution of charges, causing dipole polarization [20]. There are two obvious resonance peaks in ε'' of the S-32 sample, indicating the existence of two extremely strong polarization processes. According to the analysis above, the two resonance peaks correspond to interface polarization and dipole polarization.

In addition to dielectric loss, the introduction of Ni and Co metal particles also produces a strong magnetic loss. As shown in Figure 5d, the S-64 sample has strong resonance peaks in the low and high-frequency bands, indicating the existence of natural resonance and exchange resonance [21]. The S-64 sample has relatively low carbon content as well as high Ni content, and thus exhibits enhanced magnetic loss, which is exhibited in Figure 5f.

The microwave absorption performance is determined by two factors. One is the ability of the material to absorb microwaves, which is defined as impedance matching ($|Z_{in}/Z_0|$), and the other is the ability to attenuate microwave energy [22]. Generally speaking, the intrinsic dielectric and magnetic properties affect microwave absorption efficiency. To further evaluate the microwave absorption efficiency of materials, the attenuation constant α and impedance matching property were analyzed. Attenuation constant α can be expressed by the following equation [23]:

$$\alpha = \frac{\sqrt{2}\pi f}{c}\sqrt{(\mu''\varepsilon'' - \mu'\varepsilon') + \sqrt{(\mu''\varepsilon'' - \mu'\varepsilon')^2 + (\mu'\varepsilon'' + \mu''\varepsilon')^2}} \quad (3)$$

Figure 6a exhibits the impedance matching property of Co-Ni/C, indicating the impedance matching of S-16 and S-32 samples are relatively optimal and more microwave incident into the material. Figure 6b exhibits the attenuation constant of Co-Ni/C, indicating the S-64 sample possesses the strongest attenuation capability. Generally speaking, when $|Z_{in}/Z_0|$ equals 1, electromagnetic waves enter the material and are not reflected at the interface. Considering both impedance matching and attenuation, S-16 and S-32 have the optimal microwave absorption efficiency. On the one hand, the unique microstructure and adjustable electrical properties of Co-Ni/C contribute to impedance matching. On the other hand, metal/carbon composites have both magnetic and dielectric loss mechanisms to effectively consume incidence microwave energy. Table 1 compares the microwave absorption efficiency of prepared Co-Ni/C in this work with other reported MOF derivatives, indicating the absorber prepared in this work has excellent microwave absorption efficiency.

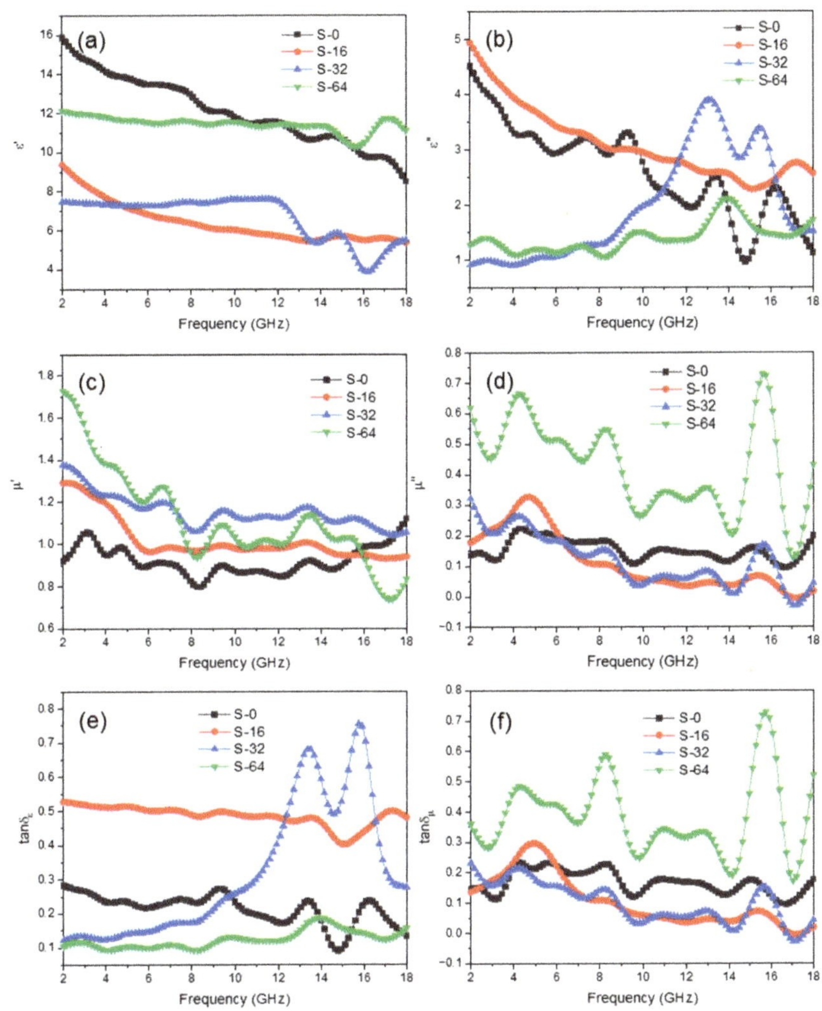

Figure 5. Electromagnetic parameters of Co-Ni/C: (**a**,**b**) real and imaginary part of permittivity, (**c**,**d**) real and imaginary part of permeability, (**e**,**f**) dielectric and magnetic loss tangent.

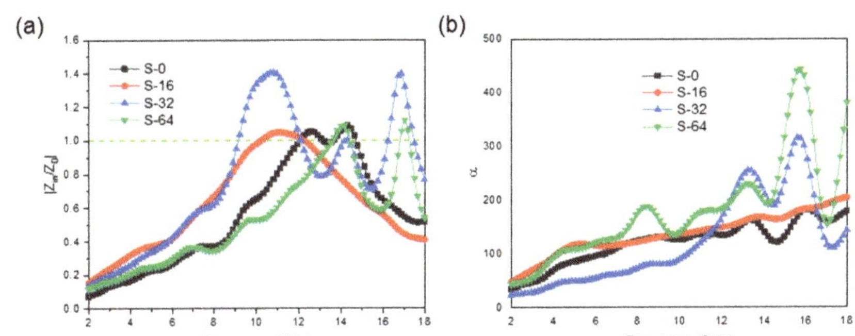

Figure 6. Impedance matching (**a**) and attenuation constant (**b**) of Co-Ni/C composites.

Table 1. Comparison of microwave absorption properties of microwave absorbers in this work with other reported MOF-derived microwave absorbers.

Sample	Absorber Thickness (mm)	RL_{max} (dB)	Effective Bandwidth (GHz)	Ref
Co/C-800	2	−32.4	3.8 (10.7–14.5)	[24]
Co/NPC@Void@CI	2.2	−49.2	6.72 (10.56–17.28)	[25]
CoS_2/NCNTs	2.5	−55	4.5 (5.5–10)	[12]
Co/C-HS-600	2	−31.3	4.4 (11.0–15.4)	[26]
$Fe_2P_4O_{12}$/phosphorus-doped carbon	2.1	−67.6	5.76 (12.24–18)	[16]
CoNi@NG-NCP	2.5	−24.03	4.32	[13]
S-16	2.7	−72.88	6.08 (11.92–18)	This work
S-32	2.21	−68.12	7.04 (9.76–16.8)	This work

4. Conclusions

In this paper, we synthesized a nano-porous Co-Ni/C composite using nano-cubic ZIF-67. After being etched by $Ni(NO_3)_2$ and in situ pyrolysis at 800 °C in Ar/H_2 atmosphere, lightweight, high-efficiency Co-Ni/C microwave absorbers were prepared. Interestingly, the Co-Ni/C composites exhibited extremely strong microwave absorption efficiency, with a RL value of −72.88 dB and, more importantly, a f_e reaching 7.04 GHz, indicating Co-Ni/C prepared in this work can be used as a high-efficiency broadband microwave absorber in stealth structure design.

Author Contributions: Data curation, Y.Z.; Formal analysis, Z.F.; Funding acquisition, Q.L.; Investigation, Q.L., Y.Z. and M.Z.; Methodology, Z.F.; Project administration, Q.L.; Resources, J.Y.; Software, Z.F.; Writing—original draft, Y.Z. and M.Z.; Writing—review & editing, J.Y. All authors have read and agreed to the published version of the manuscript.

Funding: This research received no external funding.

Institutional Review Board Statement: Not applicable.

Informed Consent Statement: Not applicable.

Data Availability Statement: All the data are available within the manuscript.

Conflicts of Interest: The authors declare no conflict of interest.

References

1. Liu, Q.; Zeng, M.; Liu, J.; Guo, B.; Hao, X.; Cao, Q.; Shang, Y.; Sun, X.; Liu, J.; Gao, X.; et al. Fe-based Material@N-doped Carbon Composites as Environment-friendly Microwave Absorbers. *Carbon* **2021**, *171*, 646–657. [CrossRef]
2. Meng, F.; Wang, H.; Huang, F.; Guo, Y.; Wang, Z.; Hui, D.; Zhou, Z. Graphene-Based Microwave Absorbing Composites: A Review and Prospective. *Compos. Part B Eng.* **2018**, *137*, 260–277. [CrossRef]
3. Cai, G.; Yan, P.; Zhang, L.; Zhou, H.; Jiang, H. Metal-Organic Framework-Based Hierarchically Porous Materials: Synthesis and Applications. *Chem. Rev.* **2021**, *121*, 12278–12326. [CrossRef]
4. Ren, S.; Yu, H.; Wang, L.; Huang, Z.; Lin, T.; Huang, Y.; Yang, J.; Hong, Y.; Liu, J. State of the Art and Prospects in Metal-Organic Frameworks-Derived Microwave Absorption Materials. *Nano Micro Lett.* **2022**, *14*, 68. [CrossRef]
5. Shu, J.; Yang, X.; Zhang, X.; Huang, X.; Cao, M.; Li, L.; Yang, H.; Cao, W. Tailoring MOF-Based Materials to Tune Electromagnetic Property for Great Microwave Absorbers and Devices. *Carbon* **2020**, *162*, 157–171. [CrossRef]
6. Zhong, G.; Liu, D.; Zhang, J. The Application of ZIF-67 and Its Derivatives: Adsorption, Separation, Electrochemistry and Catalysts. *J. Mater. Chem. A* **2018**, *6*, 1887–1899. [CrossRef]
7. Lu, Y.; Wang, Y.; Li, H.; Lin, Y.; Jiang, Z.; Xie, Z.; Kuang, Q.; Zheng, L. MOF-Derived Porous Co/C Nanocomposites with Excellent Electromagnetic Wave Absorption Properties. *ACS Appl. Mater. Interfaces* **2015**, *7*, 13604–13611. [CrossRef]
8. Huang, M.; Wang, L.; Pei, K.; You, W.; Yu, X.; Wu, Z.; Che, R. Multidimension-Controllable Synthesis of MOF-Derived Co@N-Doped Carbon Composite with Magnetic-Dielectric Synergy toward Strong Microwave Absorption. *Small* **2020**, *16*, 2000158. [CrossRef]
9. Ruan, W.; Mu, C.; Wang, B.; Nie, A.; Zhang, C.; Du, X.; Xiang, J.; Wen, F.; Liu, Z. Metal-Organic Framework Derived Cobalt Phosphosulfide with Ultrahigh Microwave Absorption Properties. *Nanotechnology* **2018**, *29*, 405703. [CrossRef]
10. Liu, P.; Gao, S.; Wang, Y.; Huang, Y.; Wang, Y.; Luo, J. Core-Shell CoNi@Graphitic Carbon Decorated on B,N-Codoped Hollow Carbon Polyhedrons toward Lightweight and High-Efficiency Microwave Attenuation. *ACS Appl. Mater. Interfaces* **2019**, *11*, 25624–25635. [CrossRef] [PubMed]

11. Liu, P.; Gao, S.; Zhang, G.; Huang, Y.; You, W.; Che, R. Hollow Engineering to Co@N-Doped Carbon Nanocages via Synergistic Protecting-Etching Strategy for Ultrahigh Microwave Absorption. *Adv. Funct. Mater.* **2021**, *31*, 2102812. [CrossRef]
12. Yan, J.; Huang, Y.; Han, X.; Gao, X.; Liu, P. Metal Organic Framework (ZIF-67)-Derived Hollow CoS_2/N-doped Carbon Nanotube Composites for Extraordinary Electromagnetic Wave Absorption. *Compos. Part B Eng.* **2019**, *163*, 67–76. [CrossRef]
13. Zhang, X.; Yan, F.; Zhang, S.; Yuan, H.; Zhu, C.; Zhang, X.; Chen, Y. Hollow N-Doped Carbon Polyhedron Containing CoNi Alloy Nanoparticles Embedded within Few-Layer N-Doped Graphene as High-Performance Electromagnetic Wave Absorbing Material. *ACS Appl. Mater. Interfaces* **2018**, *10*, 24920–24929. [CrossRef] [PubMed]
14. He, P.; Yu, X.; Lou, X. Carbon-Incorporated Nickel–Cobalt Mixed Metal Phosphide Nanoboxes with Enhanced Electrocatalytic Activity for Oxygen Evolution. *Angew. Chem. Int. Ed.* **2017**, *56*, 3897–3900. [CrossRef] [PubMed]
15. Yin, Y.; Liu, X.; Wei, X.; Yu, R.; Shui, J. Porous CNTs/Co Composite Derived from Zeolitic Imidazolate Framework: A Lightweight, Ultrathin, and Highly Efficient Electromagnetic Wave Absorber. *ACS Appl. Mater. Interfaces* **2016**, *8*, 34686–34698. [CrossRef]
16. Liu, J.; Meng, C.; Liu, Q.; Li, N.; Yu, R.; Zeng, M. Fire-resistant Iron-Based Phosphates/Phosphorus-Doped Carbon Composites Derived from Phytic Acid-Treated Metal Organic Frameworks as High-Efficiency Microwave Absorbers. *Carbon* **2022**, *200*, 472–482. [CrossRef]
17. Wang, L.; Wen, B.; Bai, X.; Liu, C.; Yang, H. Facile and Green Approach to the Synthesis of Zeolitic Imidazolate Framework Nanosheet-derived 2D Co/C Composites for a Lightweight and Highly Efficient Microwave Absorber. *J. Colloid Interface Sci.* **2019**, *540*, 30–38. [CrossRef]
18. Wang, L.; Bai, X.; Wen, B.; Du, Z.; Lin, Y. Honeycomb-Like Co/C Composites Derived from Hierarchically Nanoporous ZIF-67 as a Lightweight and Highly Efficient Microwave Absorber. *Compos. Part B: Eng.* **2019**, *166*, 464–471. [CrossRef]
19. Song, Z.; Liu, X.; Sun, X.; Li, Y.; Nie, X.; Tang, W.; Yu, R.; Shui, J. Alginate-Templated Synthesis of CoFe/carbon Fiber Composite and the Effect of Hierarchically Porous structure on Electromagnetic Wave Absorption Performance. *Carbon* **2019**, *151*, 36–45. [CrossRef]
20. Liu, G.; Liu, X.; Song, Z.; Sun, X.; Li, Y.; Shui, J.; Yu, R. Hollow Double-Shell Structured Void@SiO_2@Co-C Composite for Broadband Electromagnetic Wave Absorption. *Chem. Eng. J.* **2021**, *417*, 128093. [CrossRef]
21. Wang, H.; Xiang, L.; Wei, W.; An, J.; He, J.; Gong, C.; Hou, Y. Efficient and Lightweight Electromagnetic Wave Absorber Derived from Metal Organic Framework-Encapsulated Cobalt Nanoparticles. *ACS Appl. Mater. Interfaces* **2017**, *9*, 42102–42110. [CrossRef] [PubMed]
22. Wang, B.; Wu, Q.; Fu, Y.; Liu, T. A Review on Carbon/Magnetic Metal Composites for Microwave Absorption. *J. Mater. Sci. Technol.* **2021**, *86*, 91–109. [CrossRef]
23. Zhi, D.; Li, T.; Qi, Z.; Li, J.; Tian, Y.; Deng, W.; Meng, F. Core-Shell Heterogeneous Graphene-Based Aerogel Microspheres for High-Performance Broadband Microwave Absorption via Resonance Loss and Sequential Attenuation. *Chem. Eng. J.* **2022**, *433*, 134496. [CrossRef]
24. Qiang, R.; Du, Y.; Chen, D.; Ma, W.; Wang, Y.; Xu, P.; Ma, J.; Zhao, H.; Han, X. Electromagnetic Functionalized Co/C Composites by in Situ Pyrolysis of Metal-Organic Frameworks (ZIF-67). *J. Alloys Compd.* **2016**, *681*, 384–393. [CrossRef]
25. Quan, B.; Liang, X.; Ji, G.; Ma, J.; Ouyang, P.; Gong, H.; Xu, G.; Du, Y. Strong Electromagnetic Wave Response Derived from the Construction of Dielectric/Magnetic Media Heterostructure and Multiple Interfaces. *ACS Appl. Mater. Interfaces* **2017**, *9*, 9964–9974. [CrossRef] [PubMed]
26. Li, Z.; Han, X.; Ma, Y.; Liu, D.; Wang, Y.; Xu, P.; Li, C.; Du, Y. MOFs-Derived Hollow Co/C Microspheres with Enhanced Microwave Absorption Performance. *ACS Sustain. Chem. Eng.* **2018**, *6*, 8904–8913. [CrossRef]

coatings

MDPI

Article

Optimization of Wire EDM Process Parameters on Cutting Inconel 718 Alloy with Zinc-Diffused Coating Brass Wire Electrode Using Taguchi-DEAR Technique

Lijun Liu [1], Muthuramalingam Thangaraj [2,*], Panagiotis Karmiris-Obratański [3,*], Yuanhua Zhou [4,*], Ramamurthy Annamalai [5], Ryszard Machnik [3], Ammar Elsheikh [6] and Angelos P. Markopoulos [7]

[1] School of Mechanical Engineering, Xijing University, Xi'an 710123, China
[2] Department of Mechatronics Engineering, SRM Institute of Science and Technology, SRM Nagar, Kattankulathur 603203, India
[3] Department of Manufacturing Systems, Faculty of Mechanical Engineering and Robotics, AGH University of Science and Technology, 30-059 Cracow, Poland
[4] School of Mechanical Engineering, Yangtze University, Jingzhou 434023, China
[5] Department of Mechanical Engineering, Bharath Institute of Higher Education and Research, Tambaram, Chennai 600073, India
[6] Department of Production Engineering and Mechanical Design, Faculty of Engineering, Tanta University, Tanta 31527, Egypt
[7] Laboratory of Manufacturing Technology, School of Mechanical Engineering, National Technical University of Athens, 15780 Athens, Greece
* Correspondence: muthurat@srmist.edu.in (M.T.); karmiris@agh.edu.pl (P.K.-O.); zhouyuanhua@yangtzeu.edu.cn (Y.Z.)

Abstract: Inconel 718 alloy has a wide range of applications in the aerospace sector because of its superior mechanical properties and its weldability. The machining of such higher strength materials with complex shapes is possible with wire electrical discharge machining. In the present research, an endeavor was made to enhance the machining process by utilizing zinc-diffused coating brass wire electrode and Taguchi-Data Envelopment Analysis-based Ranking (DEAR) methodology in the process while machining Inconel 718 alloy. Material removal rate, kerf width, and surface roughness were considered as the quality measures. The optimal arrangement of input factors in the Wire Electrical Discharge Machining (WEDM) process were found as 140 µs (T_{on}), 50 µs (T_{off}), 60 V (SV), and 5 kg (WT) among the elected factors with the error accuracy of 1.1%. The pulse-off time has the most significance on formulating the quality measures owing to its importance on deionization in the process.

Keywords: Inconel 718 alloy; DWE; DEAR; optimization; WEDM

1. Introduction

Even though nickel super alloys are classified as difficult-to-cut materials, they have gained popularity based on their excellent mechanical properties in increased temperatures, including creep, fatigue, and corrosion resistance [1,2]. Thus, several industries use nickel-based alloys, including aerospace, nuclear, automotive, and chemical [3]. Nickel-based alloys such as Face Cantered Cubic (FCC) Structures are becoming increasingly important due to their ability to format Topological Close Packed (TCP) phases, which is the largest group of intermetallic compounds with complex crystallographic structures [4]. These phases increase their strength and can withstand temperatures up to 1500 °C [5]. Since nickel-based alloys are difficult-to-cut materials, the manufacturing cost of conventional machining is increased [6]. This may dramatically challenge the economic feasibility of the machining process due to their abrasive nature and the increased mechanical properties, which may lead to excessive cutting forces, tool wear, and temperatures [7]. Thus, as a cost-effective solution using unconventional machining methods such as the Wire Electrical Discharge Machining (WEDM) process, the exotic materials can be machined with the ability to produce complicated shapes [8]. WEDM removes the material by melting and vaporizing cavities using electric discharges that come from a scrolling electrical wire. During machining, the inter-electrode gap is flushed with a dielectric under a certain pressure, which removes erosion products in the machining zone, provides the appropriate conditions necessary for electrical erosion to occur, and insulates and cools the electrodes [9,10]. The dielectric is usually demineralized and deionized water, and the electrode is brass or copper wire with a zinc jacket. WEDM is a variation of EDM machining, where the main difference is the type of working electrode used. WEDM wire cutting is the most precise method of metal cutting, allowing dimensional accuracy much higher than laser cutting (thermal cutting) [11].

Electrical discharges have been established to affect the surface integrity of work pieces for many years [12,13]. Several studies have shown that the roughness of the surface of wire-EDM varies with the duration of the current pulse (T_{on}) [14–16], the peak discharge current (I_p) [17–20], and the average energy per spark [21,22]. Using a Response Surface Methodology (RSM), a mathematical model was proposed in order to evaluate the WEDM process parameters during the machining of Inconel 627 alloy [23]. Based on their results, the Material Removable Rate (MRR) can be increased up to 85% as the pulse-on time is increased, but we can achieve the same result by decreasing the pulse-off time. The dimensional accuracy of machining Inconel 718 by WEDM and producing fir tree slots were investigated [24]. By using nickel-coated electrodes, the authors tried to avoid the contamination of the workpiece and the possibility of creating intermetals. Using the composite electrodes, the machining time was reduced by 33%, but the standard brass wire electrodes produced better surface roughness. There are a lot of publications focused on the formation of the recast layer [25–28], and the surface crack density [29,30]. The formation of the recast layer on Inconel 718 using a vibration-assisted WEDM was also investigated [31]. The results revealed that within 15 to 20 μm of the wire-EDM cut surface, residual stresses changed from tensile to compressive and the recast layer thickness formed between 5 to 9 μm. Another study investigated the surface integrity and performance of WEDM cutting an Inconel 718 using coated and uncoated electrodes [32]. It was observed that by analyzing the variables, i.e., recast layer characteristics, wire feed rate, and kerf width, the zinc-coated copper electrodes have higher performance compared to the brass wire. More specifically, using the optimal machining parameters, the zinc-coated copper wire feed rate was increased by 35% while the wire consumption was reduced by 40%. Utilizing the same nickel alloy, the wire material performance in the productivity and surface integrity of the workpiece machined by WEDM was analyzed [33]. The authors have drawn several conclusions from the current study, particularly, that the discharge energy significantly influenced surface roughness and cutting speed, and when discharge energy mode was changed from the highest to the lowest, surface roughness was reduced by 46.9%. Additionally, the coating of the wire electrodes reduced the cutting time by 22%. Finally, extensive investigation of the WEDM process on precision machining of Nimonic C 236 alloy was performed [34]. The authors performed a detailed analysis of the machined surface by using Scanning Electron Microscopy (SEM) and Transmission Electron Microscopy (TEM). The results revealed that there was a diffusion process between the electrode material and the workpiece, in particular, an adhesive film of 15 wt.% Cu and 2 wt.% Zn adheres to the subsurface and surface of the sample. It was observed that zinc-diffused brass wire electrodes can produce better machinability in the WEDM process [35]. However, the implementation of optimization algorithms is important for enhancing the process mechanism [36].

As it can be seen from the above-mentioned literature review, there are limited studies that perform an optimization study of Inconel alloys and especially of the Inconel 718, by using a coated electrode. More specifically, in the present study, there is an extensive analysis of the occurred mechanisms during the machining and the relationship between the electrode conductivity and the machining performance. Although many research attempts are available on the machining of nickel-based alloy specimens, only a few research works were available to examine the influence of process control factors on quality measures while cutting this kind of alloy using the WEDM process. Only a few research endeavors were made to adopt Taguchi-DEAR as a Multi-Response Optimization Techniques (MCDT) method in the WEDM process. Hence, this experimental investigation is carried out to employ this method to optimize the WEDM cutting of chosen specimens to achieve the following goals:

- Employing Taguchi-DEAR method as a new MCDT tool to obtain optimal arrangement of process control factors of the machining process.
- Analyzing the influence of process control factors in WEDM process using different performance measures.
- To identify the most significant factors in performance measures.

2. Experiments and Methods

2.1. Selection of Workpiece Specimens and Wire Electrodes

In the present study, the utilized material was an Inconel 718 nickel-based alloy (Bibus Metals, Krakow, Poland). Inconel 718 has a wide range of applications in the aerospace sector because of its superior mechanical properties and its weldability, and it is a nickel-ferrous-chromium-based super alloy capable of withstanding temperatures up to 650 °C [37]. Due to the γ prime and γ double prime phases, the Inconel 718 alloy achieves high thermal strength [38]. Thus, this alloy is extensively used for gas turbine blades [39]. In order to achieve its strength, refractory alloying elements are added to achieve the well-known γ austenite nickel solid solution. Typical workpieces are rectangular specimens (10 mm × 10 mm × 25 mm) with a dimension of 10 mm × 10 mm and a length of 25 mm to be used as workpiece specimens.

The selection of wire electrode material plays an important role in the WEDM process. Due to their superior physical properties as depicted in Table 1, zinc-diffused brass wire electrode (DWE) was utilized as tool electrodes with a diameter of 0.25 mm. The following Table 2 contains the chemical composition of the Inconel 718 alloy using X-ray fluorescence spectroscopy (XRF) manufactured by Bruker Corporation, Massachusetts, United States. Figure 1 shows the diffused wire electrodes for machining the workpiece in the WEDM process. The brass wire electrode coated with a zinc layer of 25 μm thickness was used as coating thickness to reduce electrode wear and wire breakage [40]. The zinc-coated brass wire electrode was treated with heat for making the coated zinc to melt and attach tightly to the brass wire electrode [41]. Figure 1 shows that the zinc material was uniformly distributed around the brass wire electrode. The three-dimensional views of the machined surface wire electrode were acquired using Atomic Force Microscope (AFM), manufactured by NTEGRA PRIMA, Amsterdam, The Netherlands. The quantification of morphology and surface roughness can be crucial to formulate a quality control metric for the end product or to verify a distinct processing step. It is essentially more critical to measure surface roughness at the scale of microns. AFM has the ability to quantify three-dimensional spatial resolutions. The AFM offers total 3D surface measurement by imaging topography-based height analysis.

Table 1. Survey performed in the present study.

Properties	Unit	Conventional Electrode	Diffused Zinc Coating Brass Electrode
Melting point	°C	920	980
Tensile strength	N/mm^2	900	980

Table 2. Chemical composition (wt.%) of the 718 alloy.

C	Mn	Si	Cr	Ni	Al	Mo	Cu	Cb	Ta	Ti	Co
0.05	0.08	0.1	18.25	52.25	0.54	2.9	0.03	5.19	0.01	1.05	0.24

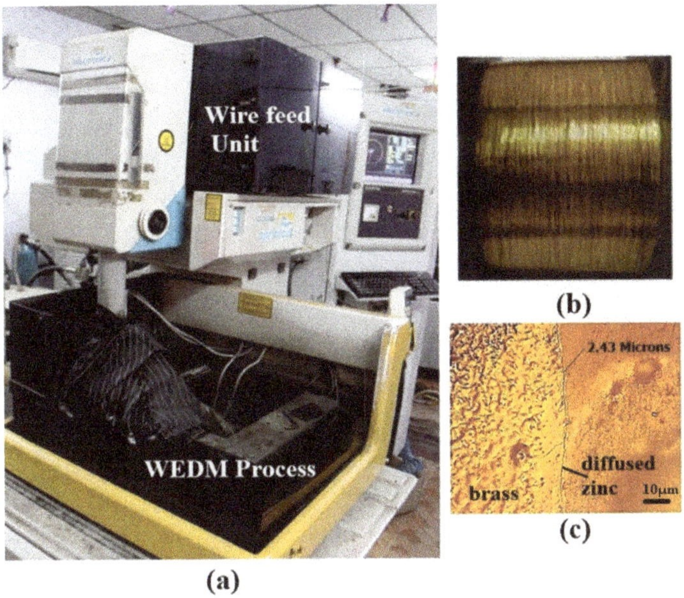

Figure 1. WEDM process arrangement: (**a**) Experimental setup; (**b**) Zinc-diffused brass wire electrode, and (**c**) Micro structure of wire electrode.

2.2. WEDM Cutting of Inconel 718 Alloy Specimens

In order to machine the specimens, an Electronica Sprintcut Electra ELPULS 40A WEDM machine (Electronica India Limted, Kolkata, India) has been used. Due to their importance on determining the machining characteristics, pulse-on time (T_{on}), pulse-off time (T_{off}), servo voltage (SV), and wire tension (WT) are chosen as the input process parameters. Pulse-on time is the duration of applying electrical energy between the work-piece and wire electrode. Since the performance measures such as material removal rate (MRR), surface roughness (R_a), and kerf width (K_w) influence the process efficiency, these parameters are taken as the performance measures of the WEDM process in the study. Deionized water was used as the insulating medium for conducting machining in the EDM process with a peak current of 16 Amps. The dielectric flow rate was chosen as 1, 2 MPa as constant throughout all the experiments with a wire feed rate as 4 m/min. The consolidated selection of process variables for this study is shown in Table 3.

Table 3. Influence of tool electrodes in the WEDM process.

Process Parameters	Unit	Variables
T_{on}	μs	100, 120, 150
T_{off}	μs	30, 40, 50
SV	V	40, 60, 80
WT	kg	5, 7, 9

In order to calculate the MRR, for each test run, the weight was measured before and after the experiment using a precision scale. It can be denoted by 3 mm/min. The

Ra of the machined specimen was evaluated by a non-contact surface roughness tester (TALYSURF CCI LITE manufactured by Taylor Hobson Ltd., Leicester, UK) with high-pass filtering based on the ISO 4287 standard. The measurement was made in the perpendicular direction to the cutting direction. The kerf width (Kw) of all machined specimens was measured using the Instron Tukon 2100B microscope from Wilson Instruments, MA, USA. Figure 2 shows the measurement approach for kerf width. It was calculated by measuring the distance between the two edge lines made by the wire passage, as shown in Figure 2. The surface topography of the machined surface was obtained using a Scanning Electron Microscope (SEM S-3400N) from Hitachi Limited, Ibaraki, Japan.

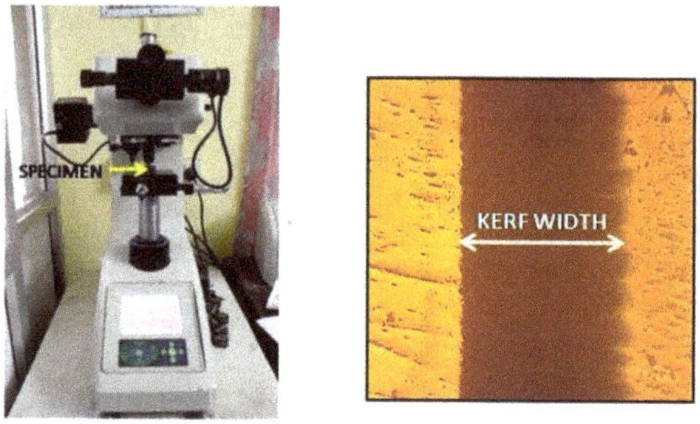

Figure 2. Measurement of K_W of machined specimen using the microscope.

2.3. *Taguchi—DEAR Technique*

Since the present study involves the four input factors with three levels and two-factor interactions between them, the L_{27} orthogonal array (OA) was selected for conducting experiments in the WEDM process according to the Taguchi technique [42]. Since the present investigation was performed with three performance measures, it was required to implement MCDT in the process.

In the proposed DEAR approach, the optimal levels of control factors can be computed using a computed ratio depending on a certain mapping process between a set of original responses and this ratio. This computed ratio is defined as the Multi-Response-Performance-Index (MRPI) value to obtain the optimal arrangement of the control factors [43,44]. Figure 3 illustrates the steps involved in the analysis.

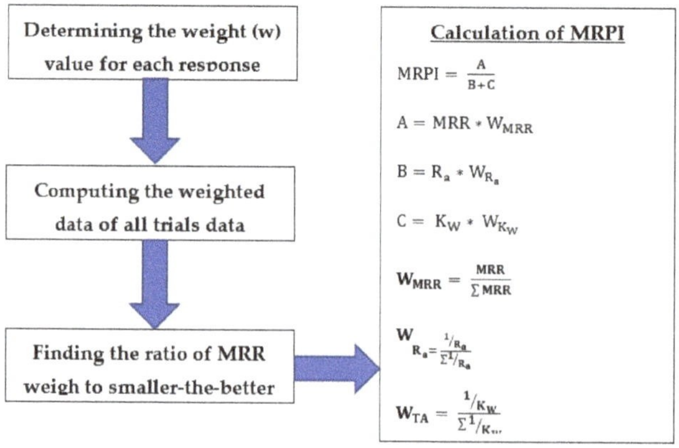

Figure 3. Steps involved in Taguchi-DEAR.

3. Results and Discussion

In the current work, 27 cutting trials were performed according to the proposed design of experiments to examine WEDM parameters' effects on workpiece specimens' machinability. Each trial was conducted three times, and the average was considered the final measured value. Table 4 lists the experimental outcome of all the conducted trials. Totally, 81 trials were performed using the WEDM process to analyze the influence of wire electrodes and process parameters on performance measures in such a process. Figure 4 shows the microstructure analysis of machined specimens using DWE in the WEDM process. The patterns of the wire electrodes are visible in specimens with a recast white-layer region, craters and an arcing region.

Table 4. Calculation of performance measures in the WEDM.

Input Factors				Quality Measures		
T_{on}	T_{off}	SV	WT	MRR (mm^3/min)	R_a (μm)	K_W (mm)
100	30	40	5	74.406	1.764	0.31
100	30	60	7	92.069	1.813	0.30
100	30	80	9	26.659	1.536	0.30
100	40	40	7	131.012	2.164	0.33
100	40	60	9	54.813	2.447	0.34
100	40	80	5	105.585	2.697	0.36
100	50	40	9	87.860	2.212	0.34
100	50	60	5	139.297	2.321	0.37
100	50	80	7	115.515	2.653	0.36
120	30	40	7	131.627	1.948	0.29
120	30	60	9	113.970	1.874	0.32
120	30	80	5	35.615	1.563	0.32
120	40	40	9	127.945	2.247	0.33
120	40	60	5	79.349	2.254	0.37
120	40	80	7	154.458	2.256	0.35
120	50	40	5	142.596	2.502	0.34
120	50	60	7	139.453	2.434	0.34
120	50	80	9	156.865	2.657	0.34
140	30	40	9	131.553	2.040	0.31
140	30	60	5	152.107	1.845	0.32
140	30	80	7	45.962	1.598	0.35
140	40	40	5	143.811	2.163	0.32
140	40	60	7	119.371	2.178	0.38
140	40	80	9	135.171	2.134	0.34
140	50	40	7	190.307	2.425	0.37
140	50	60	9	217.164	2.379	0.33
140	50	80	5	206.989	2.627	0.36

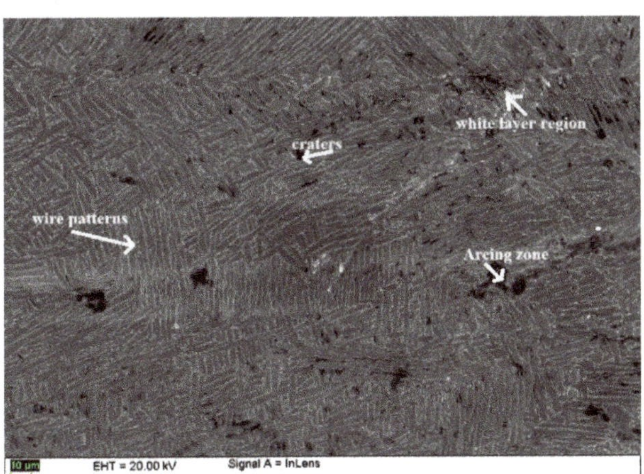

Figure 4. Machined surface of specimens in WEDM.

Figure 4 shows the microstructure analysis of machined specimens using DWE in the WEDM process. The patterns of the wire electrodes are visible with specimens with a negligible recast-white layer region, craters and an arcing region. Figure 5 shows the surface morphology of the machined specimen using the WEDM process with DWE under trial 26. It shows the peak and valley over the machined surface. The white region indicates the presence of the zinc content deliberated from the diffused wire electrode as per both 3D and 2D views of the surface. The zinc particles were deposited over the machined surface. The surface roughness can also be interpreted from the two-dimensional view of the surface. The histogram analysis shows a higher occurrence of surface roughness over the machined surface. From the analysis, it is possible to analyze an average surface roughness value from the higher occurrence of peaks and valleys of the machined surface using the WEDM process.

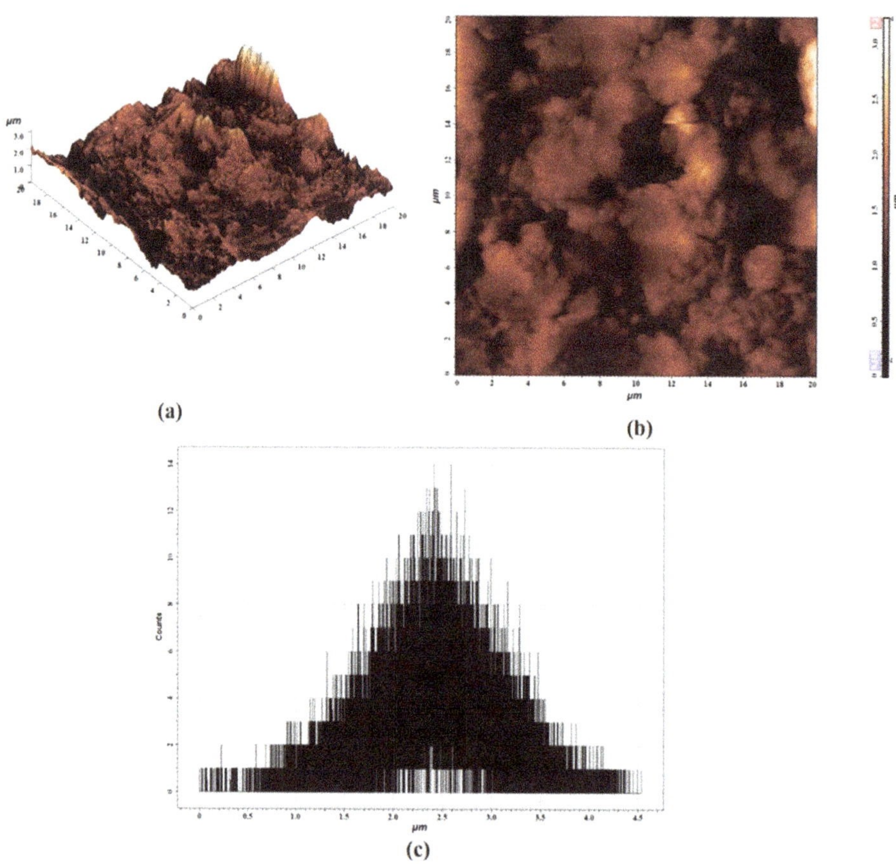

Figure 5. Surface morphology analysis of machined specimen under the factors of trial 26: (**a**) Three dimensional view; (**b**) Two dimensional view, and (**c**) Histogram analysis.

3.1. *Effects of Process Factors on Performance Measures*

Table 5 shows the weights of each outcome parameter along with the MRPI values. It was inferred that trial number 26 could remove more material with lower Ra and KW. The material removal is directly proportional to the pulse-on time. Hence, the larger pulse-on time could produce a higher MRR. Since the pulse-off time contributes for the deionization process after the machining mechanism, the larger pulse-off time could produce lower surface roughness. The servo voltage can highly contribute to the overcut [42]. The moderate servo voltage can produce lower kerf width. Hence, trial number 26 could produce a higher MRR with lower surface roughness and kerf width over the machined surface of workpiece specimens using the WEDM process with DWE.

Table 5. Calculation of weights and MRPI.

Trial No.	Weights			MRPI
	MRR	R_a	K_W	
1.	0.0229	0.0445	0.0399	18.7050
2.	0.0283	0.0433	0.0411	28.6394
3.	0.0082	0.0511	0.0411	2.4012
4.	0.0403	0.0363	0.0376	57.9910
5.	0.0169	0.0321	0.0365	10.1508
6.	0.0325	0.0291	0.0346	37.6650
7.	0.0270	0.0355	0.0365	26.0809
8.	0.0428	0.0338	0.0337	65.5571
9.	0.0355	0.0296	0.0346	45.0828
10.	0.0405	0.0403	0.0424	58.5362
11.	0.0350	0.0419	0.0387	43.8854
12.	0.0110	0.0502	0.0387	4.2856
13.	0.0393	0.0349	0.0376	55.3071
14.	0.0244	0.0348	0.0337	21.2726
15.	0.0475	0.0348	0.0355	80.6040
16.	0.0438	0.0314	0.0365	68.6989
17.	0.0429	0.0323	0.0365	65.7044
18.	0.0482	0.0295	0.0365	83.1364
19.	0.0404	0.0385	0.0399	58.4706
20.	0.0468	0.0425	0.0387	78.1695
21.	0.0141	0.0491	0.0355	7.1373
22.	0.0442	0.0363	0.0387	69.8744
23.	0.0367	0.0360	0.0329	48.1430
24.	0.0416	0.0368	0.0365	61.7309
25.	0.0585	0.0324	0.0337	122.3619
26.	0.0668	0.0330	0.0376	159.3354
27.	0.0636	0.0299	0.0346	144.7537

3.2. Computation of Significant Process Factors

In order to identify the optimal factors using the consolidated MRPI concept [43,44], another combination of input process factors may provide better performance measures than trial 26. The consolidated MRPI of all control variables with the corresponding levels is shown in Table 6. The maximum-level value of each process factor specifies the optimal level of input factors in formulating the quality measures. From Table 5, it was inferred that T_{on} (level 3), T_{off} (level 3), SV (level 2), and WT (level 1) are the optimal arrangement of process factors in the current investigation. The MRPI values with different combinations of process control variables of the WEDM process were calculated using the Taguchi-DEAR technique. The maximum value of each control variable indicates the optimal level of input control variable of any machining process based on different quality measures. It was inferred that the optimal arrangement of process control variables in the process of the current study is 140 µs (T_{on}), 50 µs (T_{off}), 60 V (SV), and 5 kg (WT) as shown in Table 7. The material removal is directly proportional to the pulse-on time. Hence, the larger pulse-on time could produce a higher MRR [45]. Since the pulse-off time contributes to the deionization process after the machining mechanism, the larger pulse-off time could produce lower surface roughness. The shape of the wire electrode can highly contribute to the overcut. The lower wire tension creates a smooth and straight wire which has approached the specimens. Hence, it can produce a lower kerf width. The highest value of max–min reveals the larger importance of process control variables on machining characteristics. It was observed that pulse-off time has higher significance in determining the quality measures due to its importance on deionization [46]. The deionization has to be performed after every pulse discharge. Then, only the next cycle of charging will be effectively performed to enhance the process mechanism. Since the flushing process performs during the pulse-off time, it can positively influence quality measures in the WEDM process.

Table 6. Computation of average MRPI values.

Factors	Levels			Max–Min
	1	2	3	
T_{on}	32.4748	53.4923	83.3307	50.8559
T_{off}	33.3589	49.1932	86.7457	53.3868
SV	49.6704	70.9961	48.6313	22.3648
WT	70.3756	51.9027	47.0196	23.3560

Table 7. Identification of optimum factors.

Factors	Variable	Unit
Pulse-on time	140	µs
Pulse-off time	50	µs
Servo voltage	60	V
Wire tension	5	kg

3.3. Confirmation of the Optimized Parameters

A confirmation trial needs to check the conclusions derived from the investigation. The experimental trial was again performed using the optimal levels of significant factors [47]. The outcome values obtained from the confirmation experiments under the computed setting are 218.234 mm^3/min (MRR), 2.345 µm (R_a), and 0.31 mm (K_W). The DEAR approach-based MRPI value of the confirmation experiment deviated by 1.1% from the mean value. Hence, the accuracy of the prediction was verified.

3.4. Surface Analysis under Optimal Control Variables

As shown in Figure 6, the surface topography was investigated using AFM under the optimal input factors in WEDM. The three-dimensional analysis showed that the lower peaks and valleys were observed over the machined surface as compared with trial 26. The histogram analysis proved that a higher occurrence of lower surface roughness points was observed over the machined surface than the surface obtained with the process factors of trial 26 in the WEDM process. Figure 7 shows the SEM-based surface morphology under optimal factors in the WEDM process. The melted particles are resolidified over the machined surface, which can be termed a white-layer thickness. The thickness of the layer can affect the surface quality of the specimens [46]. It was observed that only a little thickness of resolidified thickness was observed over the machined surface.

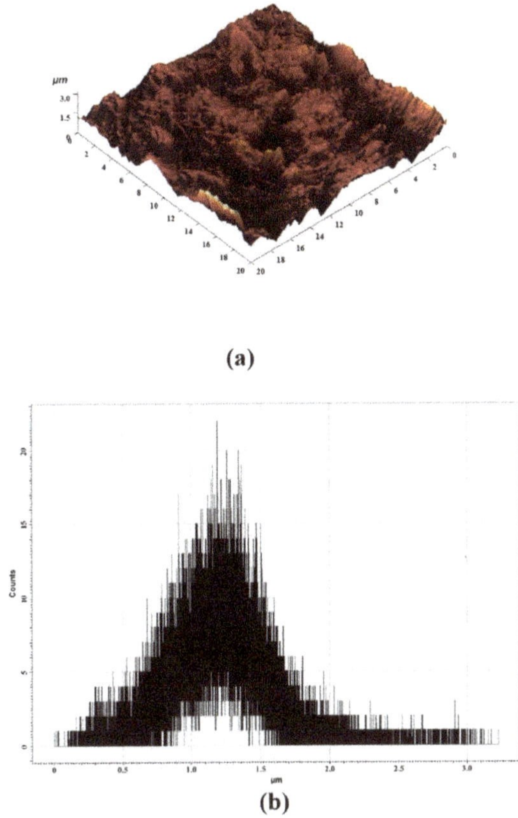

Figure 6. Surface morphology analysis of machined specimen under optimal process factors: (**a**) Three-dimensional view and (**b**) Histogram analysis.

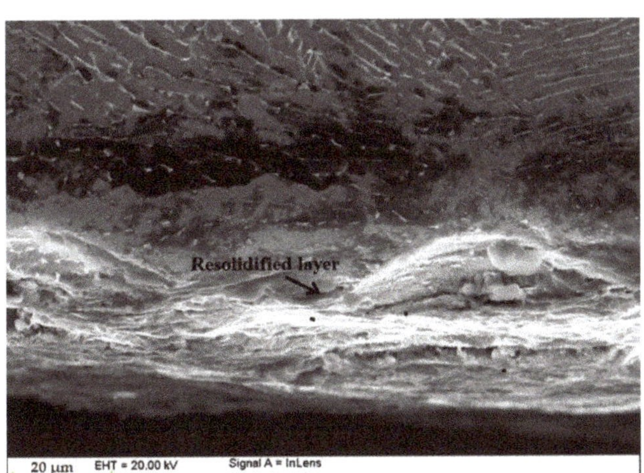

Figure 7. SEM-based surface morphology under optimal factors.

4. Conclusions

In the current investigation, an effort was made to introduce Taguchi-DEAR methodology in the WEDM process while machining Inconel 718 alloy using a zinc-diffused coating brass wire electrode. The following conclusions were drawn from it:

- ➢ The zinc-diffused coating brass wire electrode can effectively machine the Inconel 718 alloy in the process.
- ➢ The patterns of the wire electrodes are visible with specimens with negligible recast white-layer region, craters, and arcing region.
- ➢ The optimal arrangement of input factors in the WEDM process were found as 140 μs (T_{on}), 50 μs (T_{off}), 60 V (SV), and 5 kg (WT) among the elected factors with the error accuracy of 1.1%.
- ➢ The pulse-off time has the most significance on formulating the quality measures owing to its importance on deionization in the process.
- ➢ The significance of other suitable coated electrodes can be analyzed on performance measures using the WEDM process in the future.

Author Contributions: Conceptualization, M.T. and P.K.-O.; methodology, M.T. and P.K.-O.; software, M.T.; validation, L.L. and Y.Z.; formal analysis, R.A. and R.M.; investigation, M.T. and A.E.; resources, A.P.M.; writing—original draft preparation, M.T. and A.E.; project administration, M.T. and P.K.-O.; funding acquisition, M.T. and P.K.-O. All authors have read and agreed to the published version of the manuscript.

Funding: This research project was Supported By Open Fund (PLN2022-17) of the State Key Laboratory of Oil and Gas Reservoir Geology and Exploitation (Southwest Petroleum University).

Institutional Review Board Statement: Not applicable.

Informed Consent Statement: Not applicable.

Data Availability Statement: Not applicable.

Acknowledgments: This research project was partly supported by the program "Excellence initiative—research university" from the A.G.H. University of Science and Technology, grant number IDUB: 4480.

Conflicts of Interest: The authors declare no conflict of interest.

References

1. Thakur, D.G.; Ramamoorthy, B.; Vijayaraghavan, L. Study on the machinability characteristics of superalloy Inconel 718 during high speed turning. *Mater. Des.* **2009**, *30*, 1718–1725. [CrossRef]
2. Sharma, S.; Jain, V.; Shekhar, R. Electrochemical Drilling of Inconel Superalloy with Acidified Sodium Chloride Electrolyte. *Int. J. Adv. Manuf. Technol.* **2002**, *19*, 492–500. [CrossRef]
3. Chen, Y.C.; Liao, Y.S. Study on wear mechanisms in drilling of Inconel 718 superalloy. *J. Mater. Process. Technol.* **2003**, *140*, 269–273. [CrossRef]
4. Rettig, R.; Singer, R.F. Numerical modelling of precipitation of topologically close-packed phases in nickel-base superalloys. *Acta Mater.* **2011**, *59*, 317–327. [CrossRef]
5. Wilson, A.S. Formation and effect of topologically close-packed phases in nickel-base superalloys. *Mater. Sci. Technol.* **2017**, *33*, 1108–1118. [CrossRef]
6. Ezugwu, E.O. Key improvements in the machining of difficult-to-cut aerospace superalloys. *Int. J. Mach. Tools Manuf.* **2005**, *45*, 1353–1367. [CrossRef]
7. Wang, Z.Y.; Rajurkar, K.P. Cryogenic machining of hard-to-cut materials. *Wear* **2000**, *239*, 168–175. [CrossRef]
8. Kumar, V.; Jangra, K.K.; Kumar, V.; Sharma, N. WEDM of nickel based aerospace alloy: Optimization of process parameters and modelling. *Int. J. Interact. Des. Manuf.* **2017**, *11*, 917–929. [CrossRef]
9. Mahapatra, S.S.; Patnaik, A. Optimization of wire electrical discharge machining (WEDM) process parameters using Taguchi method. *Int. J. Adv. Manuf. Technol.* **2007**, *34*, 911–925. [CrossRef]
10. Kumar, A.; Kumar, V.; Kumar, J. Multi-response optimization of process parameters based on response surface methodology for pure titanium using WEDM process. *Int. J. Adv. Manuf. Technol.* **2013**, *68*, 2645–2668. [CrossRef]
11. Thangaraj, M.; Annamalai, R.; Moiduddin, K.; Alkindi, M.; Ramalingam, S.; Alghamdi, O. Enhancing the Surface Quality of Micro Titanium Alloy Specimen in WEDM Process by Adopting TGRA-Based Optimization. *Materials* **2020**, *13*, 1440. [CrossRef] [PubMed]

12. Hewidy, M.S.; El-Taweel, T.A.; El-Safty, M.F. Modelling the machining parameters of wire electrical discharge machining of Inconel 601 using RSM. *J. Mater. Process. Technol.* **2005**, *169*, 328–336. [CrossRef]
13. Newton, T.R.; Melkote, S.N.; Watkins, T.R.; Trejo, R.M.; Reister, L. Investigation of the effect of process parameters on the formation and characteristics of recast layer in wire-EDM of Inconel 718. *Mater. Sci. Eng. A.* **2009**, *513*, 208–215. [CrossRef]
14. Han, F.; Jiang, J.; Yu, D. Influence of machining parameters on surface roughness in finish cut of WEDM. *Int. J. Adv. Manuf. Technol.* **2007**, *34*, 538–546. [CrossRef]
15. Devarasiddappa, D.; Chandrasekaran, M.; Arunachalam, R. Experimental investigation and parametric optimization for minimizing surface roughness during WEDM of Ti6Al4V alloy using modified TLBO algorithm. *J. Braz. Soc. Mech. Sci. Eng.* **2020**, *42*, 128. [CrossRef]
16. Kavimani, V.; Prakash, K.S.; Thankachan, T. Multi-objective optimization in WEDM process of graphene—SiC-magnesium composite through hybrid techniques. *Measurement* **2019**, *145*, 335–349. [CrossRef]
17. Soundararajan, R.; Ramesh, A.; Mohanraj, N.; Parthasarathi, N. An investigation of material removal rate and surface roughness of squeeze casted A413 alloy on WEDM by multi response optimization using RSM. *J. Alloys Compd.* **2016**, *685*, 533–545. [CrossRef]
18. Roy, A.; Narendranath, S.; Pramanik, A. Effect of peak current and peak voltage on machined surface morphology during WEDM of TiNiCu shape memory alloys. *J. Mech. Sci. Technol.* **2020**, *34*, 3957–3961. [CrossRef]
19. Chaudhary, A.; Sharma, S.; Verma, A. Optimization of WEDM process parameters for machining of heat treated ASSAB '88 tool steel using Response surface methodology (RSM). *Mater. Today* **2022**, *50*, 917–922. [CrossRef]
20. Dabade, U.A.; Karidkar, S.S. Analysis of response variables in WEDM of Inconel 718 using Taguchi technique. *Procedia CIRP* **2016**, *41*, 886–891. [CrossRef]
21. Grigoriev, S.N.; Nadykto, A.B.; Volosova, M.A.; Zelensky, A.A.; Pivkin, P.M. WEDM as a replacement for grinding in machining ceramic Al_2O_3-TiC Cutting Inserts. *Metals* **2021**, *11*, 882. [CrossRef]
22. Maher, I.; Sarhan, A.A.D. Proposing a new performance index to identify the effect of spark energy and pulse frequency simultaneously to achieve high machining performance in WEDM. *Int. J. Adv. Manuf. Technol.* **2017**, *91*, 433–443. [CrossRef]
23. Garg, M.P.; Kumar, A.; Sahu, C.K. Mathematical modeling and analysis of WEDM machining parameters of nickel-based super alloy using response surface methodology. *Sadhana* **2017**, *42*, 981–1005. [CrossRef]
24. Klocke, F.; Welling, D.; Klink, A.; Veselovac, D.; Nöthe, T.; Perez, R. Evaluation of Advanced Wire-EDM Capabilities for the Manufacture of Fir Tree Slots in Inconel 718. *Procedia CIRP* **2014**, *14*, 430–435. [CrossRef]
25. Aspinwall, D.K.; Soo, S.L.; Berrisford, A.E.; Walder, G. Workpiece surface roughness and integrity after WEDM of Ti–6Al–4V and Inconel 718 using minimum damage generator technology. *CIRP Annals. Manuf. Technol.* **2008**, *57*, 187–190. [CrossRef]
26. Bisaria, H.; Shandilya, P. Experimental investigation on wire electric discharge machining (WEDM) of Nimonic C-263 superalloy. *Mater. Manuf. Process.* **2019**, *34*, 83–92. [CrossRef]
27. Sharma, P.; Chakradhar, D.; Narendranath, S. Analysis and Optimization of WEDM Performance Characteristics of Inconel 706 for Aerospace Application. *Silicon* **2018**, *10*, 921–930. [CrossRef]
28. Zhang, M.; Liu, Z.; Pan, H.; Deng, C.; Qiu, M. Effect of no-load rate on recast layer cutting by ultra fine wire-EDM. *Chin. J. Aeronaut.* **2021**, *34*, 124–131. [CrossRef]
29. Kumar, A.; Kumar, V.; Kumar, J. Surface crack density and recast layer thickness analysis in WEDM process through response surface methodology. *Mach. Sci. Technol.* **2016**, *20*, 201–230. [CrossRef]
30. Reddy, B.S.; Rao, A.B.K.; Janardhana, G.R. Multi-objective optimization of surface roughness, recast layer thickness and surface crack density in WEDM of Al2124/SiCp using desirability approach. *Mater. Today* **2021**, *39*, 1320–1326.
31. Zhang, Z.; Huang, H.; Ming, W.; Xu, Z.; Huang, Y.; Zhang, G. Study on machining characteristics of WEDM with ultrasonic vibration and magnetic field assisted techniques. *J. Mater. Process. Technol.* **2016**, *234*, 342–352. [CrossRef]
32. Reolon, L.W.; Laurindo, C.A.H.; Torres, R.D.; Amorim, F.L. WEDM performance and surface integrity of Inconel alloy IN718 with coated and uncoated wires. *Int. J. Adv. Manuf. Technol.* **2019**, *100*, 1981–1991. [CrossRef]
33. Abhilash, P.M.; Chakradhar, D. Effect of wire material and discharge energy on productivity and surface integrity of WEDM-processed Inconel 7180. *Adv. Mater. Process. Technol.* **2022**, 1–22. [CrossRef]
34. Mouralova, K.; Benes, L.; Bednar, J.; Zahradnicek, R.; Prokes, T.; Fiala, Z.; Fries, J. Precision Machining of Nimonic C 263 Super Alloy using WEDM. *Coatings* **2020**, *10*, 590. [CrossRef]
35. Muthuramalingam, T.; Ramamurthy, A.; Sridharan, K.; Ashwin, S. Analysis of surface performance measures on WEDM processed titanium alloy with coated electrodes. *Mater. Res. Express.* **2018**, *5*, 126503. [CrossRef]
36. Ramamurthy, A.; Sivaramakrishnan, R.; Muthuramalingam, T. Taguchi-Grey Computation Methodology for Optimum Multiple Performance Measures on machining Titanium alloy in WEDM Process. *Indian J. Eng. Mater. Sci.* **2015**, *22*, 181–186.
37. Sonar, T.; Balasubramanian, V.; Malarvizhi, S.; Venkateswaran, T.; Sivakumar, D. An overview on welding of Inconel 718 alloy—Effect of welding processes on microstructural evolution and mechanical properties of joints. *Mater. Charact.* **2021**, *174*, 110997. [CrossRef]
38. Ling, L.-S.; Yin, Z.; Hu, Z.; Liang, J.-H.; Wang, Z.-Y.; Wang, J.; Sun, B.-D. Effects of the γ''-Ni3Nb Phase on Mechanical Properties of Inconel 718 Superalloys with Different Heat Treatments. *Materials* **2019**, *13*, 151. [CrossRef]
39. Popovicha, V.A.; Borisovb, E.V.; Popovichb, A.A.; Sufiiarovb, V.S.H.; Masaylob, D.V.; Alzina, L. Impact of heat treatment on mechanical behaviour of Inconel 718 processed with tailored microstructure by selective laser melting. *Mater. Des.* **2017**, *131*, 12–22. [CrossRef]

40. Phan, N.H.; Thangaraj, M. Multi criteria decision making of vibration assisted EDM process parameters on machining silicon steel using Taguchi-DEAR methodology. *Silicon* **2021**, *13*, 1879–1885. [CrossRef]
41. Ramamurthy, A.; Sivaramakrishnan, R.; Muthuramalingam, T.; Venugopal, S. Performance analysis of wire electrodes on machining Ti-6Al-4V alloy using wire electrical discharge machining process. *Mach. Sci. Technol.* **2015**, *19*, 577–593. [CrossRef]
42. Thangaraj, M.; Ahmadein, M.; Alsaleh, N.A.; Elsheikh, A.H. Optimization of Abrasive Water Jet Machining of SiC Reinforced Aluminum Alloy Based Metal Matrix Composites Using Taguchi–DEAR Technique. *Materials* **2021**, *14*, 562–578. [CrossRef] [PubMed]
43. Phan, N.H.; Dong, P.V.; Dung, H.T.; Thien, N.V.; Muthuramalingam, T.; Shirguppikar, S.; Tam, N.C.; Ly, N.T. Multi-object optimization of EDM by Taguchi-DEAR method using AlCrNi coated electrode. *Int. J. Adv. Manuf. Technol.* **2021**, *116*, 1429–1435. [CrossRef]
44. Muthuramalingam, T.; Vasanth, S.; Vinothkumar, P.; Geethapriyan, T.; Rabik, M.M. Multi criteria decision making of abrasive flow oriented process parameters in abrasive water jet machining using Taguchi-DEAR Methodology. *Silicon* **2018**, *10*, 2015–2021. [CrossRef]
45. Phan, N.H.; Muthuramalingam, T.; Minh, N.D.; Duc, N.V. Enhancing surface morphology of machined SKD61 die steel in EDM process using DEAR approach based multi criteria decision making. *Int. J. Interact. Des. Manuf.* **2022**, *16*, 1155–1161. [CrossRef]
46. Muthuramalingam, T. Measuring the influence of discharge energy on white layer thickness in electrical discharge machining process. *Measurement* **2019**, *131*, 694–700. [CrossRef]
47. Sakthivel, G.; Saravanakumar, D.; Muthuramalingam, T. Application of failure mode and effects analysis in manufacturing industry—an integrated approach with FAHP-FUZZY TOPSIS and FAHP-FUZZY VIKOR. *Int. J. Product. Qual. Manag.* **2018**, *24*, 398–423. [CrossRef]

MDPI AG
Grosspeteranlage 5
4052 Basel
Switzerland
Tel.: +41 61 683 77 34

Coatings Editorial Office
E-mail: coatings@mdpi.com
www.mdpi.com/journal/coatings

www.ingramcontent.com/pod-product-compliance
Lightning Source LLC
Chambersburg PA
CBHW041953260926
51850CB00019B/2688
* 9 7 8 3 7 2 5 8 8 1 5 5 0 *